FE Exam Review for Mechanical Engineering

Timothy C. Kennedy, PhD, PE
Professor Emeritus of Mechanical Engineering
Oregon State University

Preface

This book provides a quick review for engineers and engineering students preparing for the Fundamentals of Engineering exam in Mechanical Engineering. The following topics are covered: Mathematics, Statistics, Computer Applications, Electrical Circuits, Statics, Mechanics of Materials, Dynamics, Systems and Controls, Materials, Machine Design, Thermodynamics, Fluid Mechanics, Heat Transfer, and Engineering Economics. This book does not aim to prepare the reader for every possible problem that might appear on the exam. Its goal is to provide a review that can be accomplished in a reasonable amount of time and still provide the reader with sufficient preparation to pass the exam.

Formulas and Units

The symbols used to represent various quantities that appear in formulas in this book were chosen to correspond to those commonly used in textbooks on a particular subject. This was done to give the reader an air of familiarity as he or she reviews each topic. Unfortunately, one of the results of this choice is that these symbols do not always match those appearing in formulas in the FE Reference Handbook. For example, the symbol $\dot{V}$ is used to represent volumetric flow rate in this book while the symbol Q is used to represent the same quantity in the Fluid Mechanics section in the FE Reference Manual. The reader is strongly encouraged to review the formulas in the FE Reference Handbook before the exam to avoid confusion.

Applying engineering formulas requires the proper use of units. The fundamental quantities occurring in Mechanical Engineering problems are time (in seconds), length (in feet or meters), and mass (in slugs or kilograms). From these, various other quantities can be expressed as follows

$$\text{Force} \rightarrow 1lb = 1slugft/s^2 \text{ or } 1Newton = 1kgm/s^2$$
$$\text{Stress} \rightarrow 1psi = 1lb/in^2 \text{ or } 1Pa = 1Newton/m^2$$
$$\text{Energy} \rightarrow 1BTU = 778ftlb \text{ or } 1J = 1Newtonm$$
$$\text{Power} \rightarrow 1hp = 550ftlb/s \text{ or } 1W = 1J/s$$

For many problems, information is provided about an object's weight rather than its mass. The weight W and mass m are related by

$$W = mg$$

where $g = 32.2ft/s^2$ or $9.81m/s^2$ is the gravitational constant.

Many problems in Thermal/Fluids engineering express mass in terms of pounds mass (lbm). This unit is related to slugs and pounds as

$$1lbm = 0.031slug \text{ and } 1lbm = 0.031lbs^2/ft$$

<u>Cautionary Note to Readers</u>

Although every effort has been made to eliminate errors in the formulas and example problems, the author acknowledges that errors may creep in through typos and other oversights. Readers may report errors to the author at <u>Timothy.Kennedy@oregonstate.edu</u>. An errata sheet can be obtained from the author at the same address.

CONTENTS

1. MATHEMATICS

1.1 Calculus

1.1.1 Differentiation

Consider a function y evaluated at x and at x+Δx as shown in Figure 1-1.

Figure 1-1

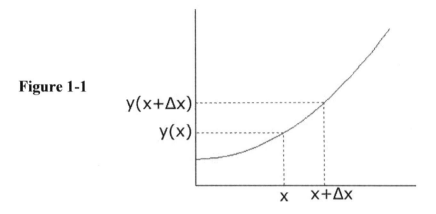

The derivative of y with respect to x is defined as

$$\frac{dy}{dx} = \lim_{\Delta x \to 0} \frac{y(x + \Delta x) - y(x)}{\Delta x}$$

From the figure, we can see that dy/dx is the slope of the curve at position x. The slope will be equal to zero at a peak (maximum) or at a trough (minimum) on the curve. The derivatives of some common functions are given in Table 1.1.

Table 1.1 Derivatives of Elementary Functions

$f(x)$	$\dfrac{df(x)}{dx}$
x^a	ax^{a-1}
e^x	e^x
$\sin x$	$\cos x$
$\cos x$	$-\sin x$

Derivatives of more complicated functions can be found by using various rules.

Chain rule: $\dfrac{dy}{dx} = \dfrac{dy}{ds}\dfrac{ds}{dx}$

Example: Find the derivative of e^{3x}.

Let s=3x.

$$\frac{d(e^{3x})}{dx} = \left[\frac{d(e^s)}{ds}\right]\left[\frac{d(3x)}{dx}\right] = [e^s][3] = 3e^{3x}$$

Example: Find the derivative of $\sin x^2$.

Let s=x².

$$\frac{d(\sin x^2)}{dx} = \left[\frac{d(\sin s)}{ds}\right]\left[\frac{d(x^2)}{dx}\right] = [\cos s][2x] = 2x\cos x^2$$

With a little practice, you should be able to do the variable substitution in your head without writing it out explicitly.

Product rule: $\dfrac{d(uv)}{dx} = u\dfrac{dv}{dx} + v\dfrac{du}{dx}$

Example: Find the derivative of $x^2 \cos x$.

$$\frac{d(x^2 \cos x)}{dx} = x^2\frac{d(\cos x)}{dx} + \cos x\frac{d(x^2)}{dx} = -x^2 \sin x + 2x\cos x$$

Higher order derivatives can be obtained by repeated application of the differentiation process.

Example: Find the second derivative of x^3.

$$\frac{d^2(x^3)}{dx^2} = \frac{d}{dx}\left[\frac{d(x^3)}{dx}\right] = \frac{d(3x^2)}{dx} = 6x$$

1.1.2 Partial derivatives

Many quantities are functions of more than one independent variable (for example, the volume of a gas will be a function of both temperature and pressure). Consider the function z(x,y)=x(1-x)y(1-y) as shown in Figure 1-2.

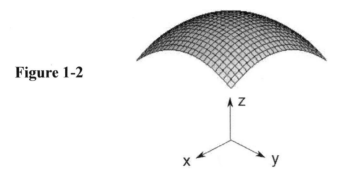

Figure 1-2

To obtain the partial derivative of z(x,y) with respect to x, we differentiate z(x,y) respect to x treating y as if it were a constant.

$$\frac{\partial z(x, y)}{\partial x} = (1 - x)y(1 - y) - xy(1 - y)$$

This quantity represents the slope of the surface while traveling along a path parallel to the x-axis. The partial derivative of z(x,y) with respect to y is

$$\frac{\partial z(x, y)}{\partial y} = x(1 - x)(1 - y) - x(1 - x)y$$

This quantity represents the slope of the surface while traveling along a path parallel to the y axis. Locating a point on the surface where there is a peak (maximum) or a trough (minimum) requires both derivatives to be zero.

1.1.3 Integration

The process that is the reverse of differentiation is called integration. The integral of the function f(x) between points x_1 and x_2 is expressed as

$$I = \int_{x_1}^{x_2} f(x)dx$$

A plot of this shown in Figure 1-3.

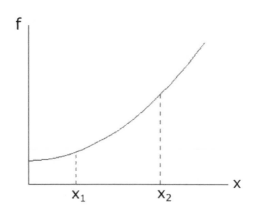

Figure 1-3

The integral of f(x) represents the area under the curve f(x) between points x_1 and x_2. Areas that dip below the x-axis are considered to be negative. When the integral is written with limits x_1 and x_2 expressed, it is called a definite integral. If the limits are omitted, it is called an indefinite integral. Indefinite integrals of some common functions are given in Table 1-2.

Table 1-2 Integrals of Elementary Functions

$f(x)$	$\int f(x)dx$
x^a	$\dfrac{x^{a+1}}{a+1}$
e^x	e^x
$\sin x$	$-\cos x$
$\cos x$	$\sin x$

In general, an integration constant should be added to each of the integrals in the table. Integrals of more complicated functions can be found with an appropriate variable substitution.

Example: Find the area under sin3x between x=0 and x=π.

$$Area = \int_0^\pi \sin 3x\, dx$$

Let s=3x. Then ds=3dx. For the new lower limit, when x=0, s=0. For the new upper limit, when x= π, s=3 π.

$$Area = \int_0^{3\pi} \sin s \left(\frac{ds}{3}\right) = -\frac{1}{3}\cos s \Big|_0^{3\pi} = -\frac{1}{3}[\cos(3\pi) - \cos(0)] = -\frac{1}{3}[-1-1] = \frac{2}{3}$$

Example: Find the area between the line y=x and the curve y=x² between the limits x=0 and x=1 as shown in Figure 1-4.

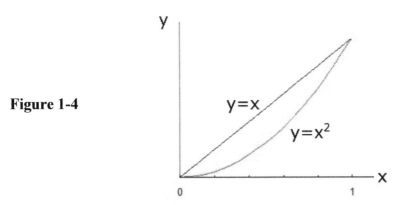

Figure 1-4

Ignoring the curve y=x², we can find the area under the line y=x as the area of the triangle $A_1=(1)(1)/2=0.5$. The area under the curve y=x² can be found by integration.

$$A_2 = \int_0^1 x^2 dx = \frac{x^3}{3}\bigg|_0^1 = \frac{1}{3}$$

The area between them is Area=A_1-A_2=0.5-0.333=0.167.

1.1.4 Numerical Integration

It is not uncommon to have a situation where the function f(x) is too complicated to evaluate in closed form. We can estimate the value of an integral by estimating the area under the curve f(x). Consider the situation shown in Figure 1-5.

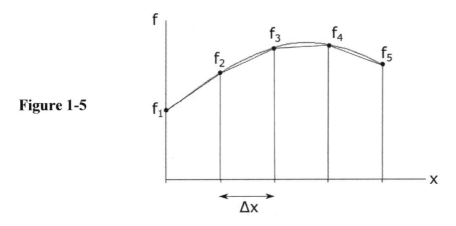

Figure 1-5

Suppose we divide the x-axis into four equal segments of length Δx and evaluate the function f(x) at the five points indicated in the figure. Next, we connect the points with straight lines. Below each of these straight lines is a trapezoid. The area contained in the trapezoid is easily calculated. For example, the area inside the first trapezoid is

$$A_1 = f_1 \Delta x + \frac{1}{2}(f_2 - f_1)\Delta x$$

Adding up the areas in each trapezoid will give an approximation of the area under the curve. From the figure, it is clear that the accuracy of the estimate will increase as the number of segments into which the curve is divided is increased. This technique is called the trapezoidal rule.

1.2 Differential Equations

As the name implies, differential equations are equations containing derivatives of some unknown function y(x). The order of the equation will equal the order of the highest derivative. For example, an equation where d^3y/dx^3 is the highest order derivative will be a third order equation. The complete solution requires a specification of the boundary conditions that go along with the equation. The number of boundary conditions should equal the order of the equation. The technique for solving an equation will vary depending on the equation type. Here, we will consider the common situation of constant coefficients. If the terms that multiply y and all of its derivatives are constants, we look for a solution in the form of an exponential (i.e., $y=Ce^{rx}$).

Example: Find a solution of

$$\frac{dy}{dx} = -2y \qquad \text{with} \quad y(0) = 10$$

To solve this, we let $y = Ce^{rx}$ and substitute this into the equation.

$$\frac{d(Ce^{rx})}{dx} = -2(Ce^{rx})$$
$$rCe^{rx} = -2Ce^{rx}$$

Solving r for gives r=-2. Then,

$$y = Ce^{-2x}$$

To find C, we use the boundary condition that y must equal 10 when x=0; i.e.

$$10 = Ce^{-2(0)}$$

Solving for C gives $C = 10$. Therefore, the final solution is

$$y = 10e^{-2x}$$

Many physical processes are governed by an equation of the form

$$\frac{d^2y}{dx^2} + a\frac{dy}{dx} + by = 0$$

Setting $y = Ce^{rx}$ and substituting gives

$$r^2Ce^{rx} + arCe^{rx} + bCe^{rx} = 0$$

Eliminating Ce^{rx} in each term gives the characteristic equation for r as

$$r^2 + ar + b = 0$$

Using the quadratic formula to solve this gives the two roots

$$r_{1,2} = \frac{-a \pm \sqrt{a^2 - 4b}}{2}$$

Depending on the values of a and b, three possibly cases arise:

Case 1 $a^2 > 4b$: The roots are real ($r = r_1, r_2$). Then,

$$y = C_1e^{r_1x} + C_2e^{r_2x}$$

Case 2 $a^2 = 4b$: There are two equal roots ($r = r_1, r_1$). Then,

$$y = (C_1 + C_2x)e^{r_1x}$$

Case 3 $a^2 < 4b$: The roots are complex ($r_1 = \alpha + j\beta$, $r_1 = \alpha - j\beta$). Then,

$$y = e^{\alpha x}(C_1\cos\beta x + C_2\sin\beta x) \qquad \text{where} \quad \alpha = \frac{-a}{2} \quad \text{and} \quad \beta = \frac{\sqrt{4b - a^2}}{2}$$

Example: Find the solution to

$$\frac{d^2y}{dx^2} + 2\frac{dy}{dx} + 5y = 0 \quad \text{with} \quad y(0) = 1 \quad \text{and} \quad \frac{dy(0)}{dx} = 0$$

In this case a=2 and b=5. Therefore $a^2 < 4b$, and we have case 3 with

$$\alpha = \frac{-a}{2} = \frac{-2}{2} = -1 \quad \text{and} \quad \beta = \frac{\sqrt{4b - a^2}}{2} = \frac{\sqrt{4(5) - (2)^2}}{2} = 2$$

Then,

$$y = e^{-x}(C_1 \cos 2x + C_2 \sin 2x)$$

Find $\dfrac{dy}{dx}$ as

$$\frac{dy}{dx} = -e^{-x}(C_1 \cos 2x + C_2 \sin 2x) + e^{-x}(-2C_1 \sin 2x + 2C_2 \sin 2x)$$

Apply the boundary conditions

$$y(0) = C_1 = 1 \quad \rightarrow \quad C_1 = 1$$
$$\frac{dy(0)}{dx} = -C_1 + 2C_2 = 0 \quad \rightarrow \quad C_2 = 0.5$$

The final solution is

$$y = e^{-x}(\cos 2x + 0.5\sin 2x)$$

1.3 Vectors

Consider a point 1 in space with coordinates (x_1, y_1, z_1) as shown in Figure 1-6.

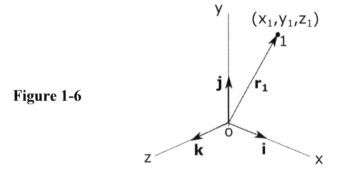

Figure 1-6

Using the unit vectors **i**, **j**, and **k**, we can construct a vector running from the origin at point o to point 1 giving

$$\mathbf{r}_1 = x_1\mathbf{i} + y_1\mathbf{j} + z_1\mathbf{k}$$

This vector has magnitude or length given by

$$L = |\mathbf{r}_1| = \sqrt{x_1^2 + y_1^2 + z_1^2}$$

and direction given by

$$\theta_x = \cos^{-1} x_1 / L \qquad \theta_y = \cos^{-1} y_1 / L \qquad \theta_z = \cos^{-1} z_1 / L$$

where θ_x, θ_y, and θ_z are the angles that the vector makes relative to the x, y, and z-axes, respectively. A unit vector (i.e., a vector with length 1) in the direction of this vector is given by

$$\mathbf{u}_1 = \frac{x_1}{L}\mathbf{i} + \frac{y_1}{L}\mathbf{j} + \frac{z_1}{L_1}\mathbf{k}$$

Many operations with vectors can be given a geometric interpretation. These operations are more easily visualized in two dimensions. Hence, we will limit further discussion to vectors with x and y components only.

1.3.1 Vector addition

Many physical quantities (e.g. force, velocity, acceleration, etc.) have both magnitude and direction. Consequently, they can be conveniently represented by vectors. Consider two forces applied at o as shown in Figure 1-7.

Figure 1-7 $\qquad \mathbf{F_2} = F_{2x}\mathbf{i} + F_{2y}\mathbf{j} \qquad \mathbf{F_1} = F_{1x}\mathbf{i} + F_{1y}\mathbf{j}$

O

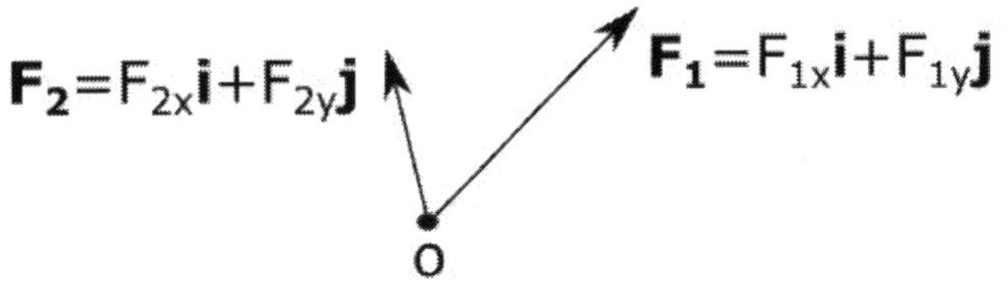

To obtain the resulting force at point o, we simply add these two vectors together

$$\mathbf{F}_r = \mathbf{F}_1 + \mathbf{F}_2 = (F_{1x} + F_{2x})\mathbf{i} + (F_{1y} + F_{2y})\mathbf{j}$$

This resultant force is simply the diagonal of the parallelogram formed by the two vectors as shown in Figure 1-8.

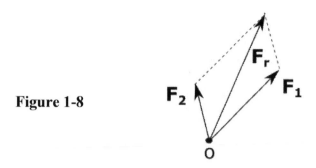

Figure 1-8

1.3.2 Vector multiplication

There are two varieties of vector multiplication. The dot product (or scalar product) of two vectors **a** and **b** is a scaler c given by

$$c = (a_x\mathbf{i} + a_y\mathbf{j} + a_z\mathbf{k}) \bullet (b_x\mathbf{i} + b_y\mathbf{j} + b_z\mathbf{k}) = a_xb_x + a_yb_y + a_zb_z = |\mathbf{a}||\mathbf{b}|\cos\theta$$

where θ is the angle between **a** and **b**.

The cross product (or vector product) of two vectors **a** and **b** is a vector **c** given by

$$\mathbf{c} = (a_x\mathbf{i} + a_y\mathbf{j} + a_z\mathbf{k}) \times (b_x\mathbf{i} + b_y\mathbf{j} + b_z\mathbf{k})$$
$$\mathbf{c} = (a_yb_z - a_zb_y)\mathbf{i} + (a_zb_x - a_xb_z)\mathbf{j} + (a_xb_y - a_yb_x)\mathbf{k} = |a||b|\mathbf{n}\sin\theta$$

where **n** is a unit vector perpendicular to the plane containing **a** and **b**, and θ is the angle measured from **a** to **b**. This operation follows the right-hand rule; i.e., imagine curling your fingers on your right hand to rotate vector **a** onto vector **b** with your thumb pointing straight up. Vector **c** will be in the direction of your thumb.

Example: Find the moment about point o caused by the force $\mathbf{F_1}$ at point 1 with coordinates (x_1,y_1) on the body shown in Figure 1-9.

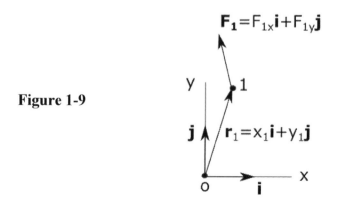

Figure 1-9

The moment about o is defined as

$$\mathbf{M} = \mathbf{r}_1 \times \mathbf{F}_1 = (x_1\mathbf{i} + y_1\mathbf{j}) \times (F_{1x}\mathbf{i} + F_{1y}\mathbf{j}) = (x_1 F_{1y} - y_1 F_{1x})\mathbf{k}$$

This moment points in the z-direction. From $|\mathbf{M}| = |\mathbf{r}_1||\mathbf{F}_1|\sin\theta$, we see from Figure 1-10

Figure 1-10

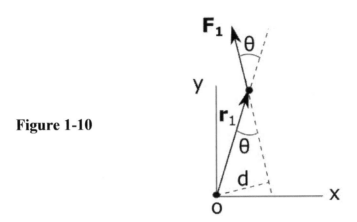

that $d = |\mathbf{r}_1|\sin\theta$ is the perpendicular distance from point o to the line of action of $\mathbf{F}_1$.

2. STATISTICS

2.1 Probability

Suppose there are n possible outcomes of an event, each with equal probability. For example, tossing a die has six possible outcomes: 1, 2, 3, 4, 5, or 6. Each has a probability of 1/6. The probability of a toss coming up 2 is 1/6. The probability of a toss coming up a 2 or a 5 is 1/6+1/6=1/3. For the "either or" situation, the probabilities are additive.

If two dice are tossed, the probability of both dice being 2 is (1/6)(1/6)=1/36. For the "both" situation, the probabilities are multiplicative. Another way to look at this is the observation that if two dice are tossed, there are 36 possible outcomes. Therefore, the outcome with both dice coming up 2 is one of 36 possible outcomes.

Example: A jar contains a large number of beans. One third of them are white, and two thirds of them are black. If two beans are withdrawn, what is the probability that we end up with one white bean and one black bean?

The probability of withdrawing a white bean is 1/3, and the probability of drawing a black bean is 2/3. The probability of drawing a white bean followed by a black bean is

$$P_1 = (1/3)(2/3) = 2/9$$

The probability of drawing a black bean followed by a white bean is

$$P_2 = (2/3)(1/3) = 2/9$$

The probability of either the first outcome or the second outcome is

$$P_{either} = P_1 + P_2 = 2/9 + 2/9 = 4/9$$

Another way to look at this is as follows. The probability of both withdrawals being white is

$$P_1 = (1/3)(1/3) = 1/9$$

The probability of both withdrawals being black is

$$P_2 = (2/3)(2/3) = 4/9$$

The probability of either of these is

$$P_{either} = P_1 + P_2 = 1/9 + 4/9 = 5/9$$

The probability that neither occurs is

$$P_{neither} = 1 - P_{either} = 1 - 5/9 = 4/9$$

2.2 Statistical Analysis

For computation purposes, it is convenient to represent probability using a probability density function f(x) where the probability that X takes on a value between a and b is

$$P(a \leq X \leq b) = \int_a^b f(x)dx$$

A commonly used probability density function is the normal (Gaussian) distribution

$$f(x) = \frac{1}{\sigma\sqrt{2\pi}} e^{-\frac{1}{2}\left(\frac{x-\mu}{\sigma}\right)^2}$$

where μ is the mean (average) defined as

$$\mu = \frac{1}{n}(X_1 + X_2 + \dots + X_n) = \frac{1}{n}\sum_i^n X_i$$

and σ is the standard deviation defined as

$$\sigma = \left[\frac{1}{n}[(X_1 - \mu)^2 + (X_2 - \mu)^2 + \dots + (X_n - \mu)^2]\right]^{1/2} = \left[\frac{1}{n}\sum_i^n (X_i - \mu)^2\right]^{1/2}$$

and X_1, X_2, … X_n are values of a random sample of n observations. The function takes on the shape of a bell with a narrow-shaped bell for small σ and a wide-shaped bell for large σ. The probability that X takes on a value between a and b is

$$P(a \leq X \leq b) = \int_a^b \frac{1}{\sigma\sqrt{2\pi}} e^{-\frac{1}{2}\left(\frac{x-\mu}{\sigma}\right)^2} dx$$

It is not possible to evaluate this integral in closed form. However, we can make a tabulated version of this if we make the change of variable

$$z = \frac{x - \mu}{\sigma}$$

Then $dz = \dfrac{dx}{\sigma}$

The new integration limits are as follows:

Lower limit of integration $z = \dfrac{a - \mu}{\sigma}$ when x=a

Upper limit of integration $z = \dfrac{b - \mu}{\sigma}$ when x=b

The probability X of being less than or equal to X_1 is

$$P(X \le X_1) = \int_{-\infty}^{\frac{X_1-\mu}{\sigma}} \frac{1}{\sqrt{2\pi}} e^{-\frac{1}{2}z^2} dz = F\left(\frac{X_1 - \mu}{\sigma}\right)$$

The probability X of being greater than or equal to X_2 is

$$P(X \ge X_2) = \int_{\frac{X_2-\mu}{\sigma}}^{\infty} \frac{1}{\sqrt{2\pi}} e^{-\frac{1}{2}z^2} dz = R\left(\frac{X_2 - \mu}{\sigma}\right)$$

Note that F(-x)=R(x).

The probability of X being in the range $X_1 \le X \le X_2$ is

$$P(X_1 \le X \le X_2) = 1 - \left[F\left(\frac{X_1 - \mu}{\sigma}\right) + R\left(\frac{X_2 - \mu}{\sigma}\right) \right]$$

For the case where $X_1=\mu-\varepsilon$ and $X_2=\mu+\varepsilon$, we have

$$P(\mu - \varepsilon \le X \le \mu + \varepsilon) = \int_{-\frac{\varepsilon}{\sigma}}^{\frac{\varepsilon}{\sigma}} \frac{1}{\sqrt{2\pi}} e^{-\frac{1}{2}z^2} dz = 1 - \left[F\left(\frac{-\varepsilon}{\sigma}\right) + R\left(\frac{\varepsilon}{\sigma}\right) \right] = 1 - 2R\left(\frac{\varepsilon}{\sigma}\right) = W\left(\frac{\varepsilon}{\sigma}\right)$$

Values of F, R, and W are given in Table 2-1. This is similar to the table on page 46 in the FE Reference Handbook.

Table 2-1   Normal Distribution

x	f(x)	F(x)	R(x)	2R(x)	W(x)
0	0.3989	0.5	0.5	1	0
0.1	0.397	0.5398	0.4602	0.9203	0.0797
0.2	0.391	5793	0.4207	0.8415	0.1585
0.3	0.3814	0.6179	0.3821	0.7642	0.2358
0.4	0.3683	0.6554	0.3446	0.6892	0.3108
0.5	0.3521	0.6915	0.3085	0.6171	0.3829
0.6	0.3332	0.7257	0.2743	0.5485	0.4515
0.7	0.3123	0.758	0.242	0.4839	0.5161
0.8	0.2897	0.7881	0.2119	0.4237	0.5763
0.9	0.2661	0.8159	0.1841	0.3681	0.6319
1	0.242	0.8413	0.1587	0.3173	0.6827
1.1	0.2179	0.8643	0.357	0.2713	0.7287
1.2	0.1942	0.8849	0.1151	0.2301	0.7699
1.3	0.1714	0.9032	0.0968	0.1936	0.8064
1.4	0.1497	0.9192	0.0808	0.1615	0.8385
1.5	0.1295	0.9332	0.0668	0.1336	0.8664
1.6	0.1109	0.9452	0.0548	0.1096	0.8904
1.7	0.094	0.9554	0.0446	0.0891	0.9109
1.8	0.079	0.9641	0.0359	0.0719	0.9281
1.9	0.0656	0.9713	0.0287	0.0574	0.9426
2	0.054	0.9772	0.0228	0.0455	0.9545
2.1	0.044	0.9821	0.0179	0.0357	0.9643
2.2	0.0355	0.9861	0.0139	0.0278	0.9722
2.3	0.0283	0.9893	0.0107	0.0214	0.9786
2.4	0.0224	0.9918	0.0082	0.0164	0.9836
2.5	0.0175	0.9938	0.0062	0.0124	0.9876
2.6	0.0136	0.9953	0.0447	0.0093	0.9907
2.7	0.0104	0.9965	0.0035	0.0069	0.9931
2.8	0.0079	0.9974	0.0026	0.0051	0.9949
2.9	0.006	0.9981	0.0019	0.0037	0.9963
3	0.0044	0.9987	0.0013	0.0027	0.9973

Example: A machine in a lumber mill is designed to cut boards to a nominal length of 96in. To evaluate the machine's performance, a sample of 1000 boards was taken from the machine, and the length of each was measured within a precision of 0.01in. The measured lengths range from 95.90in to 96.10in. The results of all the measurements are given in Table 2.2. Assuming that this sample provides an indication of future performance of the machine, find the probability of the following:

a) a board will have a length less than or equal to 95.97in.
b) a board will have a length greater than or equal to 96.05in.
c) a board will have a length between 95.98in and 96.02in

Table 2.2 Board Measurement Data

Number of Boards	Length (in)
2	95.9
4	95.91
8	95.92
15	95.93
25	95.94
39	95.95
57	95.96
77	95.97
96	95.98
109	95.99
115	96
112	96.01
97	96.02
81	96.03
61	96.04
43	96.05
28	96.06
16	96.07
9	96.08
4	96.09
2	96.1

First, we calculate the mean.

$$\mu = [2(95.9in) + 4(95.91in) + ... + 2(96.1in)]/1000 = 96.00in$$

Next, we calculate the standard deviation.

$$\sigma = \{[2(95.9in - 96in)^2 + 4(95.91in - 96in)^2 + ... + 2(96.1in - 96in)^2]/1000\}^{1/2} = 0.0345in$$

The probability that a board will have a length less than or equal to 95.97in is

$$P(X \leq 95.97in) = F\left(\frac{95.97in - 96.00in}{0.0345in}\right) = F(-0.87)$$

The table does not have values of F for negative values of x. Therefore, we use the relation F(-x)=R(x). We need to do a linear interpolation on the table between 0.8 and 0.9 giving

$$F(-0.87) = R(0.87) = 0.2119 - (0.2119 - 0.1841)(0.7) = 0.192$$

The probability that a board will have a length greater than or equal to 96.05in is

$$P(X \geq 96.05in) = R\left(\frac{96.05in - 96.00in}{0.0345in}\right) = R(1.45) = 0.0808 - (0.0808 - 0.0688)(0.5) = 0.0748$$

The probability that a board will have a length between 95.98in and 96.02in is

$$P(95.98in \leq X \leq 96.02in) = W\left(\frac{96.02in - 96.00in}{0.0345in}\right) = W(0.58) = 0.3829 + (0.4515 - 0.3829)(0.8) = 0.438$$

3. COMPUTER APPLICATIONS

3.1 Flow Charts

A flowchart is a diagram showing the logic involved in performing a computation on a computer. Typically, it will involve repeated application of a formula containing an equal sign. The use of an equal sign in a computer algorithm is different from the usual mathematical definition. Each variable in a computation has a unique storage location in the computer. The name of the variable and the storage location are synonymous. The equal sign tells the computer to replace what is stored on the left side of the equation with whatever appears on the right side. For example, the equation $x=x+1$ makes no sense from a mathematical standpoint. However, to a computer it says to replace what is stored in x by the value x plus 1.

Example: Given an integer N, find the sum of the squares of all the integers running from 1 to N.

The logic flowchart for this calculation for N=5 is shown in Figure 3-1.

Figure 3-1

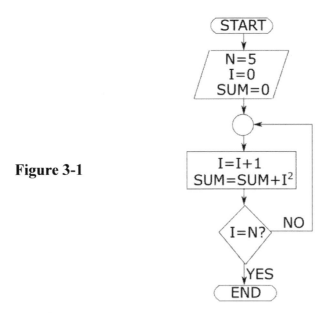

From the flowchart, we observe that the calculation of adding the current integer's squared value to the current sum loops back until we reach the final integer as N. The sequence for calculating the value of SUM will go as follows:

$$I=1 \quad SUM=0+1^2=1$$
$$I=2 \quad SUM=1+2^2=5$$
$$I=3 \quad SUM=5+3^2=14$$
etc

3.2 Spreadsheets

A spreadsheet is a computer application useful for calculations involving tabular data. It has the capability of performing repetitious calculations without the need for using a computer language to write a computer program. The spreadsheet is composed of cells with each cell being identified by a column designation (in terms of letters of the alphabet) and a row designation (in terms of numbers). For example, cell B3 is in column B and row 3 as shown in Figure 3-2.

Figure 3-2

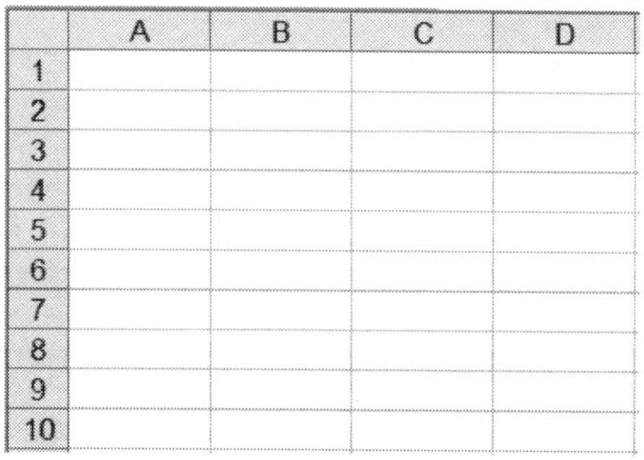

	A	B	C	D
1				
2				
3				
4				
5				
6				
7				
8				
9				
10				

A calculation can be performed by entering a formula into a cell. The variables of the formula are identified according to their cell location. For example, if "=A1+B1" is typed into cell C1, the values in cells A1 and B1 are added together and entered into C1. If there were values in cells A1 through A10 and B1 through B10, we could copy the formula in C1 to cells C2 through C10. In this case C2 would contain A2+B2, cell C3 would contain A3+B3, etc. By writing the formula as "=A1+B1", the cells A1 and B1 are given "relative" references in this formula. This means that the cell row numbers in columns A and B will automatically be incremented as we copy the formula into rows 2 through 10 in column C. If we copied the formula in cell C1 to cell D1, then cell D1 would contain cells B1 and C1 added together. In this case by copying the cell from column C to column D in row 1, the column designations are automatically incremented.

We can prevent automatic incrementation in the formula copying process by using an "absolute" reference rather than a relative reference. We can make the column designation absolute by putting a $ symbol in front of the column letter. This prevents column letter incrementation during the copying process. If we had typed the formula in C1 as "=$A1+B1", then we would get the sum of A1 and C1 in cell D1 when we copied the formula to cell D1. We would get the sum of A1 and D1 when we copied the formula to cell E1.

We can make the row designation absolute by putting the $ symbol in front of the row number. This prevents the row numbers from being automatically incremented during the copying process. If we had written the formula as "A$1+B1" in cell C1, we get the sum of A1+B2 in cell C2 when we copy the formula to cell C2. Next, we would get the sum of A1+B3 when we copy the formula to cell C3. Putting the $ symbol in front of both the column letter and the row number prevents both from being incremented.

Example: Create a spreadsheet that calculates the probability density function for a normal distribution

$$f(x) = \frac{1}{\sigma\sqrt{2\pi}} e^{-\frac{1}{2}\left(\frac{x-\mu}{\sigma}\right)^2}$$

over ten increments of x between X_1 and X_2 for values of μ and σ.

We will do this for X_1=95.9, X_2=96.1, μ=96.0 and σ=0.0345.

We begin by entering the given values 95.9 into cell A1, 96.1 into B1, 96.0 into C1, and 0.0345 into D1. Next, we enter the formula "=($B$1-$A$1)/10" into cell A2 to calculate the increment size for x. Then, we enter $A$1 into cell A4 as the starting point for x. Next, we enter the formula "=A4+$A$2" into cell A5 as the first increment in x. Then, we copy cell A5 into cells A6 through A14. Now, cells A4 through A14 contain the range in x from 95.9 to 96.1. Next, we enter the formula

=1/[$D$1*SQRT(2*3.1416)]*EXP(-0.5*((A4-$C$1)/$D$1)^2))

into cell B4 to get f(x) for the value of x contained in cell A4. Then, we copy cell B4 into cells B5 to B14. Now, cells B4 through B14 contain values of f(x) for the values of x in cells A4 to A14 as shown in Figure 3-3.

Figure 3-3

	A	B	C	D	E
1	95.9	96.1	96.0	0.0345	
2	0.02				
3					
4	95.90	0.0001			
5	95.92	0.0018			
6	95.94	0.0194			
7	95.96	0.1040			
8	95.98	0.2851			
9	96.00	0.3989			
10	96.02	0.2851			
11	96.04	0.1040			
12	96.06	0.0194			
13	96.08	0.0018			
14	96.10	0.0001			
15					

4. ELECTRICAL CIRCUITS

4.1 DC Circuits

Direct current circuits typically contain resistors and constant voltage sources. The relationship between voltage drop V across the resistor and current I transmitted through it is

$$V = IR$$

where V is in volts, I is in amperes, and R is in ohms. For resistors connected in series, the equivalent resistance is

$$R_{eq} = R_1 + R_2 + ... + R_n$$

For resistors connected in parallel, the equivalent resistance is

$$R_{eq} = \frac{1}{\dfrac{1}{R_1} + \dfrac{1}{R_2} + ... \dfrac{1}{R_n}}$$

The power P dissipated through resistor is given by

$$P = IV = I^2 R$$

<u>Example</u>: Find the current in the voltage source in the circuit shown in Figure 4-1.

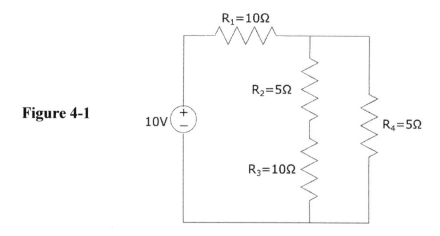

Figure 4-1

First, we combine resistors R_2 and R_3 to find the equivalent resistance R_{23eq} as

$$R_{23eq} = R_2 + R_3 = 5\Omega + 10\Omega = 15\Omega$$

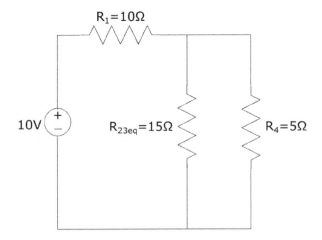

Next, we combine resistors R_{23eq} and R_4 to find the equivalent resistance R_{234eq} as

$$R_{234eq} = \frac{1}{\dfrac{1}{R_{23eq}} + \dfrac{1}{R_4}} = \frac{1}{\dfrac{1}{15\Omega} + \dfrac{1}{5\Omega}} = 3.75\Omega$$

Next, we combine resistors R_{234eq} and R_1 to find the equivalent resistance R_{1234eq} as

$$R_{1234eq} = R_1 + R_{234eq} = 10\Omega + 3.75\Omega = 13.75\Omega$$

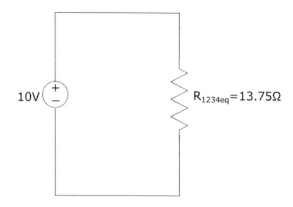

The current is then

$$I = \frac{V}{R_{1234eq}} = \frac{10V}{13.75\Omega} = 0.727\,A$$

Circuits with multiple voltage sources can be analyzing using Kirchhoff's current and voltage laws. Kirchhoff's current law states that the total current leaving a node (joint) must equal the total current entering the node; i. e.,

$$\sum I_{out} = \sum I_{in}$$

Kirchhoff's voltage law states that for any closed loop within a circuit, the sum of the voltage drops across the resistors and voltage gains across the sources must be zero; i.e.,

$$\sum V_{gain} - \sum V_{drop} = 0$$

Circuits can be analyzed by using the following steps:
1) assign a current label and direction for each loop in the circuit.
2) Apply Kirchhoff's voltage law to each loop.
3) Solve for the currents in each loop from the equations in step two.

Example: Find the current through the 20V voltage source in Figure 4-2.

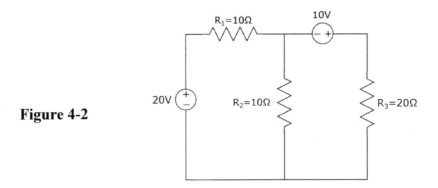

Figure 4-2

We assign a current I_A running clockwise in the left-hand loop and a current I_B running clockwise in the right-hand loop as shown in Figure 4-3.

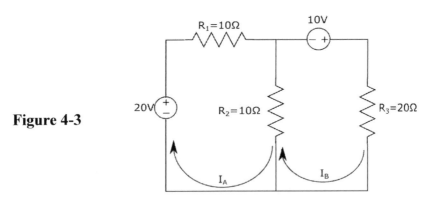

Figure 4-3

If we have guessed the wrong direction for a current, it will come out negative in the solution. The current passing through resistor R_1 is I_A. For resistor R_2, we can think of current I_A running down and current I_B running up, resulting in a net current of I_A-I_B. The current passing through resistor R_3 is I_B.

Applying Kirchhoff's voltage law to left-hand loop gives

$$20V - 10\Omega(I_A) - 10\Omega(I_A - I_B) = 0$$

Applying Kirchhoff's voltage law to the right-hand loop gives

$$10V - 20\Omega(I_B) + 10\Omega(I_A - I_B) = 0$$

Solving the two equations for I_A gives the current through the voltage source as

$$I_A = 1.4A$$

4.2 AC Circuits

Alternating current circuits typically contain voltage sources that fluctuate over time with frequency ω (in radians per second), which results in a current flow with the same frequency. This can be represented mathematically as

$$V = A\sin(\omega t + \varphi)$$

In addition to resistors, these circuits often contain capacitors and inductors. A capacitor is a component that stores energy in the form of a charge separation. The relationship between current and voltage for a capacitor is

$$I = C\frac{dV}{dt}$$

where C is the capacitance in farads, V is in volts, and I is in amperes. An inductor is a component that stores energy in a magnetic field. The relationship between current and voltage for an inductor is

$$V = L\frac{dI}{dt}$$

where L is the inductance is in henrys, V is in volts, and I is in amperes.

Rather than using a sine function, it is convenient to express AC voltage using a complex exponential

$$V = Ae^{j\omega t}$$

where $j = \sqrt{-1}$. With this we can express the impedance Z of each component as a generalized resistance by considering the ratio of voltage to current.

$$Z(\omega) = \frac{V(\omega)}{I(\omega)}$$

For a resistor $\qquad Z_R = R$

For a capacitor $\qquad Z_C = \dfrac{1}{j\omega C} = \dfrac{-j}{\omega C}$

For an inductor $\qquad Z_L = j\omega L$

Example: Find the impedance for 1) a resistor and a capacitor in series, and 2) a resistor and an inductor in series in a circuit with a 60 Hz voltage source as shown in Figure 4-4.

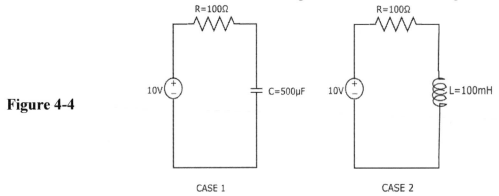

Figure 4-4

CASE 1   CASE 2

First, we convert the frequency from Hz to radians per second.

$$\omega = 60Hz\left(\frac{2\pi rad/s}{Hz}\right) = 377s^{-1}$$

Case 1:

$$Z_R = 100 \qquad \text{and} \qquad Z_C = \frac{-j}{\omega C} = \frac{-j}{(377)5x10^{-4}} = -5.31j$$

$$Z_{eq} = Z_R + Z_C = 100 - 5.31j$$

Case 2:

$$Z_R = 100 \qquad \text{and} \qquad Z_L = j\omega L = j(377)(100x10^{-3}) = 37.7j$$

$$Z_{eq} = Z_R + Z_L = 100 + 37.7j$$

To express power in an AC circuit, we first need to define a root mean square voltage as

$$V_{rms} = \sqrt{\frac{1}{T}\int_0^T V^2(t)dt}$$

For a sinusoidal voltage

$$V_{rms} = V_{max}/\sqrt{2}$$

We have a similar definition of root mean square current. The average power can then be expressed as

$$P_A = V_{rms}I_{rms}\cos\theta$$

where θ is the phase angle between the voltage and the current. The cosθ term is also called the power factor.

Example: Find the average power consumed in an AC circuit with a peak voltage of 240V, a peak current of 2A, and a phase angle of 20° between voltage and current.

$$P_A = V_{rms} I_{rms} \cos \theta = \left(\frac{V_{max}}{\sqrt{2}} \right) \left(\frac{I_{max}}{\sqrt{2}} \right) \cos \theta = \left(\frac{240V}{\sqrt{2}} \right) \left(\frac{2A}{\sqrt{2}} \right) \cos 20° = 226W$$

5. STATICS

Statics deals with the analysis of stationary rigid bodies subjected to various types of loading conditions. Internal loads and support reactions are determined using Newton's second and third laws. For stationary bodies, Newton's second law states that a body is in equilibrium if the net force and net moment acting on the body are zero; i.e.,

$$\mathbf{F}_{net} = 0 \qquad \mathbf{M}_{net=0}$$

These can be written in component form as

$$\sum F_x = 0 \qquad \sum F_y = 0 \qquad \sum F_z = 0$$

$$\sum M_x = 0 \qquad \sum M_y = 0 \qquad \sum M_z = 0$$

As described in Section 1.3.2, the moment about point o of a force acting at point 1 is equal to the cross product of the position vector from point o to point 1 and the force vector; i.e.,

$$\mathbf{M}_o = \mathbf{r}_1 \times \mathbf{F}_1$$

Alternatively, the magnitude of the moment about o is equal to the magnitude of the force times the perpendicular distance d to the line of action of the force as shown in Figure 5-1.

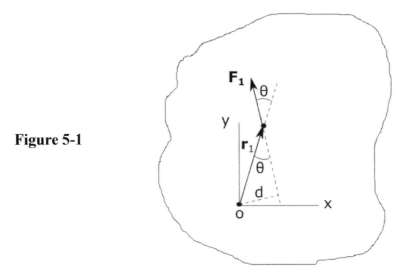

Figure 5-1

$$M_o = F_1 d$$

The direction of the moment is perpendicular to the plane containing $\mathbf{r}_1$ and $\mathbf{F}_1$ and follows the right-hand rule.

The correct application of Newton's second law requires the construction of a free body diagram. A free body diagram shows a sketch of the body or a part of the body indicating all the forces and moments that the body experiences after it has been cut free from the attachment points to its surroundings. The reaction loads that a body experiences at the attachment points depend on how the attachment inhibits movement of that point on the body. A support that inhibits displacement in a certain direction generally produces a reaction force in that direction. Similarly, a support that inhibits rotation about a certain axis would generally produce a reaction moment in that direction. Common cases are shown below.

Fixed Support (welded connection)

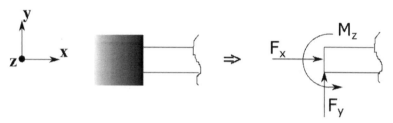

Simple Support without Sliding (stationary hinge)

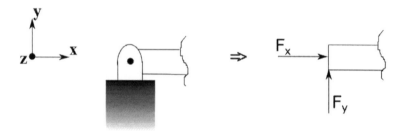

Simple Support with Sliding (sliding hinge)

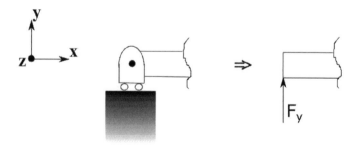

As an example, let's consider the beam shown in Figure 5-2.

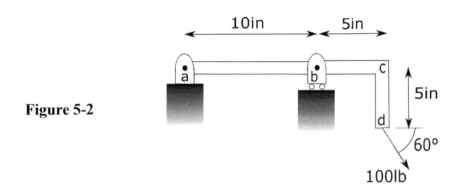

Figure 5-2

To find the reaction forces on the beam from the supports, we draw a free body diagram of the beam with the supports removed as shown in Figure 5-3.

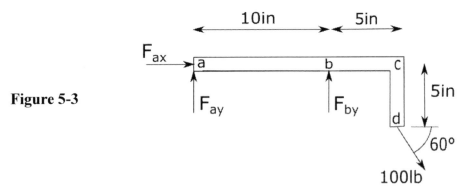

Figure 5-3

The forces and moments in the free body diagram are entered into the equilibrium equations giving

$$\sum F_x = 0 \rightarrow F_{ax} + 100lb(\cos 60°) = 0$$

$$\sum F_y = 0 \rightarrow F_{ay} + F_{by} - 100lb(\sin 60°) = 0$$

$$\sum M_{z-pointa} = 0 \rightarrow F_{by}(10in) + [(100lb)(\cos 60°)]5in - [(100lb)(\sin 60°)]15in = 0$$

We have three equations with three unknowns which can be solved to give

$$F_{ax} = -50lb \qquad F_{ay} = -18.4lb \qquad F_{by} = 105lb$$

The fact that the forces F_{ax} and F_{ay} came out negative indicates that they act opposite to the way they were drawn on the free body diagram.

If we wanted to find the internal forces and moments and at point e (8in to the right of point a), we would cut the beam at e and draw a free body diagram of the left-hand portion as shown in Figure 5-4. Since the left side of the beam is effectively welded to the right side of the beam at point e, we apply reaction loads at e in the free body diagram that are consistent with a fixed support.

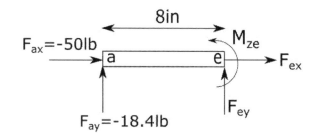

Figure 5-4

Equilibrium requires

$$\sum F_x = 0 \rightarrow -50lb + F_{ex} = 0$$
$$\sum F_y = 0 \rightarrow -18.4lb + F_{ey} = 0$$
$$\sum M_{z-pointe} = 0 \rightarrow M_{ze} - (-18.4lb)(8in) = 0$$

Solving gives

$$F_{ex} = 50lb \qquad F_{ey} = 18.4lb \qquad M_{ze} = -147inlb$$

These results indicate that the beam experiences an internal axial force of 50lb, a transverse shear force of 18.4lb , and a bending moment of 147inlb at point e. These results are needed to determine the stress in the beam at point e.

5.1 Distributed Loads

Distributed loads are typically in the form of a force per unit length. They can be handled by replacing them with statically equivalent point loads which are equal to the total force and located at the centroid of the load distribution. The two most common cases are shown below.

Uniform Load

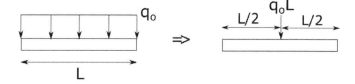

Linearly Varying Load

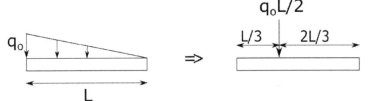

Example: Find the reaction forces at the supports for the beam in Figure 5-5.

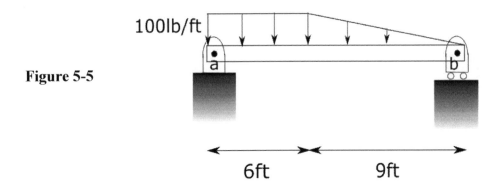

Figure 5-5

First, we replace the distributed loads with statically equivalent point loads, and then we draw a free body diagram of the beam as shown in Figure 5-6.

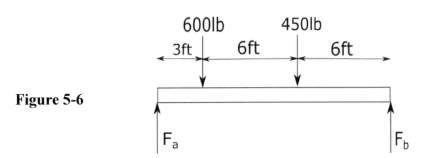

Figure 5-6

Equilibrium requires

$$\sum F_y = 0 \rightarrow F_a + F_b - 600lb - 450lb = 0$$
$$\sum M_{z-pointb} = 0 \rightarrow -F_a(15\,ft) + 600lb(12\,ft) + 450lb(6\,ft) = 0$$

Solving gives

$$F_a = 660lb \qquad\qquad F_b = 390lb$$

5.2 Truss Structures

Truss structures are composed of pin connected bars arranged in triangular patterns. All of the external loads are applied at the joints, and all supports are of the hinge type. This results in each bar carrying axial load only (no shear or bending). The analysis of these structures follows the same methods described earlier.

<u>Example:</u> Find the reaction forces at the supports for the truss structure shown in Figure 5-7.

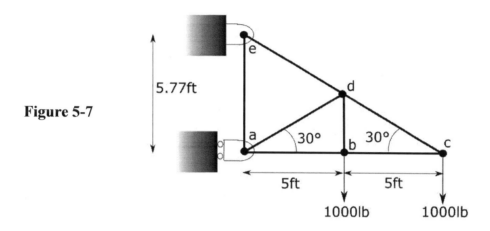

Figure 5-7

We begin by drawing a free body diagram of the entire structure separated from its supports as shown in Figure 5-8.

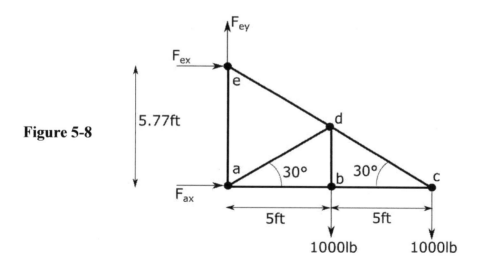

Figure 5-8

Equilibrium requires

$$\sum F_x = 0 \rightarrow F_{ax} + F_{ex} = 0$$
$$\sum F_y = 0 \rightarrow F_{ey} - 1000lb - 1000lb = 0$$

$$\sum M_{z-pointe} = 0 \rightarrow F_{ax}(5.77\,ft) - 1000lb(5\,ft) - 1000lb(10\,ft) = 0$$

Solving gives

$$F_{ax} = 2600lb \quad F_{ex} = -2600lb \quad F_{ey} = 2000lb$$

To find the internal forces in individual bars, we can use the method of joints. In this method, we imagine each bar having a hole at its ends with pins in each of these holes connecting the bars together. Let's draw a free body diagram of the pin at joint e. This pin will experience the support forces F_{ex} and F_{ey}. It will also feel the forces from bars ea and ed. Since these bars carry axial load only, the forces on the pin from bars ea and ed will be directed along the length of these bars as shown in Figure 5-9.

Figure 5-9

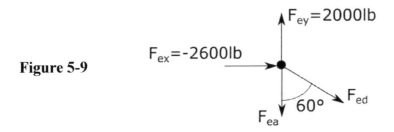

Applying force equilibrium gives

$$\sum F_x = 0 \rightarrow -2600lb + F_{ed}(\sin 60°) = 0$$
$$\sum F_y = 0 \rightarrow 2000lb - F_{ea} - F_{ed}(\cos 60°) = 0$$

Solving gives

$$F_{ea} = 500lb \quad F_{ed} = 3000lb$$

We could calculate the other internal forces by continuing this process of drawing free body diagrams of each pin and applying force equilibrium.

Suppose we are only interested in the internal force in bar ad. In this case we could apply the method of sections which involves cutting the structure through the bar of interest and drawing a free body of a section of the structure as shown in Figure 5-10.

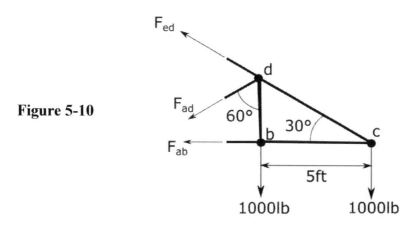

Figure 5-10

We could proceed by applying force and moment equilibrium. However, since we are only interested in bar ad, we could proceed directly to applying moment equilibrium about point c (which has the advantage of not involving the forces F_{ed} and F_{ab}) giving

$$\sum M_{z-pointc} = 0 \rightarrow (F_{ad} \cos 60°)(5\,ft) + (F_{ad} \sin 60°)(2.89\,ft) + 1000lb(5\,ft) = 0$$
$$F_{ad} = -1000lb$$

The negative sign for F_{ad} indicates that it is a compressive force.

5.2.1 Zero force members

Occasionally, truss structures will contain bars that carry no load. These are called zero force members. There is a case where these members can be identified without performing an analysis. If three bars meet at a joint without external load and two of these bars are collinear, then the third bar has zero force. For example, consider again the truss structure in Figure 5-7. Supposed the external load of 1000lb at joint b did not exist. Then bars ab and bc are collinear, and bar bd must be a zero force member. With bar bd carrying no load, we can pretend that this bar does not exist. With this development, we can identify bar ad as a zero force member. Keep in mind that this is only true if the external load at joint b does not exist.

5.3 Friction

5.3.1 Straight surfaces

Friction force occurs at the contact surface between two bodies undergoing forces that tend to make the bodies slide over one another. The magnitude of this force ranges from zero up to a maximum of μN (which occurs when one of the bodies is on the verge of slipping). N is the normal force at the contact surface, and μ is the coefficient of friction. Thus, we have

$$F_{friction} \leq \mu N$$

Example: A block weighing 100lb rests on top of another block weighing 100lb as shown in Figure 5-11. The top block is tethered to the wall that so that it cannot move. The coefficient of friction for all services is μ=0.2. Find the maximum force P that can be applied to the lower block without having it slide.

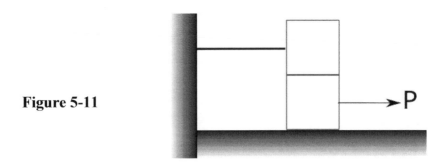

Figure 5-11

We begin by drawing free body diagrams of each block as shown in Figure 5-12.

Figure 5-12

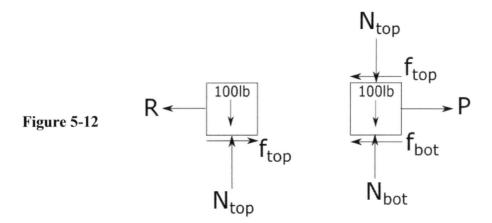

From force equilibrium in the y-direction, we can see that N_{top}=100lb and N_{bot}=200lb. When the bottom block is on the verge of sliding f_{top}= μ N_{top} and f_{bot}= μ N_{bot}. Applying force equilibrium in the x-direction on the bottom block gives

$$\sum F_x = 0 \rightarrow P - 0.2(100lb) - 0.2(200lb) = 0$$

Therefore, $P = 60lb$

5.3.2 Curved surfaces

When the contact surface between two bodies is curved, the direction of the friction force follows the contour of the surface. We will deal with the case where the surface is a portion of a circle. Consider a rope wrapped around a tree trunk with force F_1 at one end and F_2 at the other end as shown in Figure 5-13.

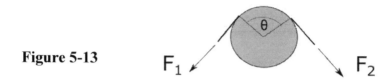

Figure 5-13

The contact angle around the circle is θ in radians. If the direction of slip is in the F_1 direction, the relationship between the forces is

$$F_1 = F_2 e^{\mu\theta}$$

<u>Example:</u> A 200lb climber dangles from a rope wrapped partially around a tree limb as shown in Figure 5-14. Find the force applied by his partner to keep him from falling; i.e., to keep the rope from sliding down if μ=0.4.

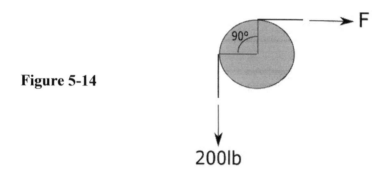

Figure 5-14

In this case, F_1=200lb and θ=90°=π/2.

$$200lb = Fe^{0.4(\pi/2)}$$

From this we get

$$F = 107lb$$

Now, suppose that the climber needs to be pulled up by his partner; i.e., the rope slides up. In this case F_2=200lb, and we get

$$F = 200lbe^{0.4(\pi/2)}$$
$$F = 375lb$$

6. MECHANICS OF MATERIALS

6.1 Stress and Strain

Mechanics of materials deals with stresses and strains in stationary, deformable bodies. The state of stress at a point in a body can be represented pictorially on a small cube around the point as shown in Figure 6-1. On each face of the cube, there are one normal stress represented by the symbol σ and two shear stresses represented by the symbol τ. The normal stresses act perpendicular to the faces of the cube causing it to stretch or compress resulting in normal strains represented by the symbol ε. The shear stresses act tangent to the faces of the cube causing it to distort resulting in shear strains represented by the symbol γ. The stresses have units of force per unit area. The normal strains are in terms of elongation per unit length. The shear strains represent the change in angle between the faces of the cube and have units of radians; for example, γ_{xy} represents the change in angle between the x and y-axes on the edges of the cube.

Figure 6-1

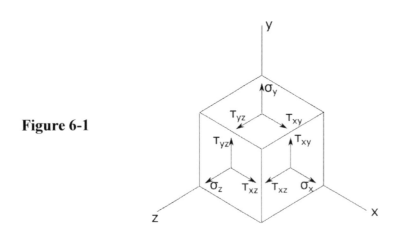

The stresses and strains are related by Hooke's law as follows:

$$\varepsilon_x = \frac{1}{E}[\sigma_x - v(\sigma_y + \sigma_z)] \qquad \gamma_{xy} = \frac{\tau_{xy}}{G}$$

$$\varepsilon_y = \frac{1}{E}[\sigma_y - v(\sigma_x + \sigma_z)] \qquad \gamma_{xz} = \frac{\tau_{xz}}{G} \qquad G = \frac{E}{2(1+v)}$$

$$\varepsilon_z = \frac{1}{E}[\sigma_z - v(\sigma_x + \sigma_y)] \qquad \gamma_{yz} = \frac{\tau_{yz}}{G}$$

where E is Young's modulus, v is Poisson's ratio, and G is the shear modulus, which are material properties. If a temperature change occurs, a term $\alpha\Delta T$ (where α is the coefficient of thermal expansion) can be added to each of the equations for normal strain.

The free surface of the body has zero stress on that surface. For a material point on the surface, we can imagine that the cube shown in Figure 6-1 has the z-face on the surface. In this case, all the stresses on that face (σ_z, τ_{xz}, τ_{yz}) are zero. The stress-strain equations for this case (plane stress) become

$$\varepsilon_x = \frac{1}{E}[\sigma_x - v\sigma_y]$$

$$\varepsilon_y = \frac{1}{E}[\sigma_y - v\sigma_x]$$

$$\gamma_{xy} = \frac{\tau_{xy}}{G} = \frac{2(1+v)}{E}\tau_{xy}$$

6.2 Principal Stress

For the plane stress case, suppose we have a different set of axes x_1, y_1 that make an angle θ relative to the original axes x,y as shown in Figure 6-2.

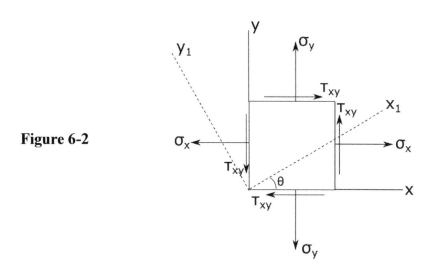

Figure 6-2

The stresses for these new directions are

$$\sigma_{x_1} = \frac{\sigma_x + \sigma_y}{2} + \frac{\sigma_x - \sigma_y}{2}\cos 2\theta + \tau_{xy}\sin 2\theta$$

$$\sigma_{y_1} = \frac{\sigma_x + \sigma_y}{2} - \frac{(\sigma_x - \sigma_y)}{2}\cos 2\theta - \tau_{xy}\sin 2\theta$$

$$\tau_{x_1 y_1} = -\frac{(\sigma_x - \sigma_y)}{2}\sin 2\theta + \tau_{xy}\cos 2\theta$$

The direction that gives the maximum (or minimum) normal stress in the x,y-plane is given by

$$\theta_p = \tan^{-1}\left(\frac{2\tau_{xy}}{\sigma_x - \sigma_y}\right)$$

where

$$\sigma_{max,min} = \frac{\sigma_x + \sigma_y}{2} \pm \sqrt{\frac{(\sigma_x - \sigma_y)^2}{4} + \tau_{xy}^2}$$

These are the principal stresses. The angle that gives the maximum shear stress in the x,y-plane is

$$\theta_s = \theta_p + 45°$$

where

$$\tau_{max} = \frac{\sigma_{max} - \sigma_{min}}{2}$$

Example: For the plane stress state shown in Figure 6-3, find the maximum shear stress in the x,y-plane.

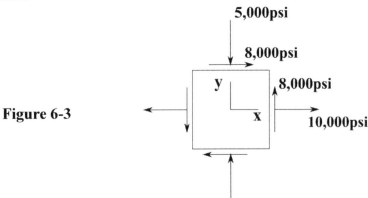

Figure 6-3

$$\sigma_{max,min} = \frac{10,000\,psi - 5,000\,psi}{2} \pm \sqrt{\frac{(10,000\,psi + 5,000\,psi)^2}{4} + (8,000\,psi)^2} = 13,466\,psi; -8,466\,psi$$

$$\tau_{max} = \frac{13,466\,psi - (-8,466\,psi)}{2} = 10,966\,psi$$

6.3 Principle of Superposition

The principle of superposition states that the response of a body to multiple loads can be obtained by adding up the responses of the body caused by each load acting individually. This will allow us to convert a complex problem into a series of simpler problems. We will use this principle to analyze any structure subjected to more than one load.

6.4 Uniaxial Loading of a Bar

Consider a bar under a load acting along the axis of the bar as shown in Figure 6-4.

Figure 6-4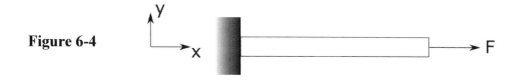

The only stress in the bar is a uniform normal stress in the x-direction.

$$\sigma_x = \frac{F}{A}$$

where A is the cross-sectional area of the bar. The strain in the bar in the x-direction is

$$\varepsilon_x = \frac{\delta}{L}$$

where δ is the stretch of the bar. From Hooke's law, the strain in the x-direction is

$$\varepsilon_x = \frac{\sigma_x}{E} = \frac{1}{E}\frac{F}{A}$$

The Poisson effect will cause the bar to contract in the y and z- directions according to

$$\varepsilon_y = \varepsilon_z = \frac{-\nu\sigma_x}{E} = \frac{-\nu}{E}\frac{F}{A}$$

From these relations, we can express the stretch of the bar in terms of the force F as

$$\delta = \frac{FL}{EA}$$

6.4.1 Non-uniform bar

Consider a bar with properties E_1 and A_1 in section 1 between points a and b, and properties E_2 and A_2 in section 2 between points b and c as shown in Figure 6-5.

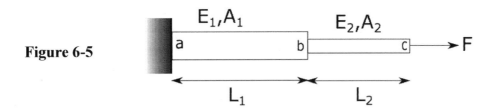

Figure 6-5

We can analyze the bar by considering each section individually. Section 1 experiences an internal force F which results in

$$\sigma_1 = \frac{F}{A_1} \qquad \delta_1 = \frac{FL_1}{E_1 A_1}$$

where σ_1 is the stress in section 1, and δ_1 is the stretch of section 1 (i.e., the amount that point b displaces relative to point a). Section 2 experiences an internal force equal to F which results in

$$\sigma_2 = \frac{F}{A_2} \qquad \delta_2 = \frac{FL_2}{E_2 A_2}$$

where σ_2 is the stress in section 2, and δ_2 is the stretch of section 2 (i.e., the amount that point c displaces relative to point b). The total stretch of the bar (the amount that point c displaces relative to point a) is

$$\delta = \delta_1 + \delta_2 = F\left(\frac{L_1}{E_1 A_1} + \frac{L_2}{E_2 A_2} \right)$$

If a bar is subjected to multiple loads, we will analyze the bar with each load applied individually and add up the results.

Example: Find the stress in each section and the displacement of the right end of the bar shown in Figure 6-6.

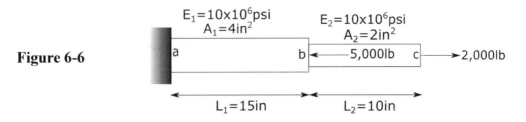

Figure 6-6

We will consider the loads one at a time.

<u>5,000lb Load Effect</u>: Consider the 5,000lb load acting alone as shown in Figure 6-7.

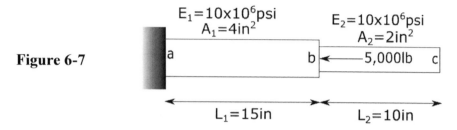

Figure 6-7

Section 1 experiences a 5,000lb internal compressive load resulting in

$$\sigma_1 = \frac{-5,000lb}{4in^2} = -1,250\,psi \qquad\qquad \delta_1 = \frac{-5000lb(15in)}{10\times10^6\,psi(4in^2)} = -0.001875in$$

We recognize that section 2 has zero internal load which means zero stress and zero stretch.

$$\sigma_2 = 0 \qquad\qquad \delta_2 = 0$$

<u>2,000lb Load Effect</u>: We now consider the 2,000lb load acting alone as shown in Figure 6-8.

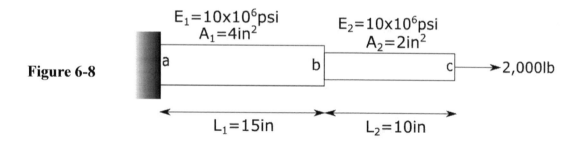

Figure 6-8

Section 1 experiences a 2,000lb internal tension force resulting in

$$\sigma_1 = \frac{2,000lb}{4in^2} = 500\,psi \qquad\qquad \delta_1 = \frac{2,000lb(15in)}{10\times10^6\,psi(4in^2)} = 0.00075in$$

Section 2 experiences a 2,000lb internal tension force resulting in

$$\sigma_2 = \frac{2,000lb}{2in^2} = 1,000\,psi \qquad\qquad \delta_2 = \frac{2,000lb(10in)}{10\times10^6\,psi(2in^2)} = 0.001in$$

<u>Combined Load Effect</u>: We now add up the individual responses.

$$\sigma_1 == -1,250\,psi + 500\,psi = -750\,psi$$

$$\sigma_2 = 0 + 1,000\,psi = 1,000\,psi$$
$$\delta_1 = -0.001875in + 0.00075in = -0.001125in$$
$$\delta_2 = 0 + 0.001in = 0.001in$$

The total stretch of the bar is
$$\delta = \delta_1 + \delta_2 = -0.001125in + 0.001in = -0.000125in$$

6.4.2 Thermal effects

If a bar that is free to expand undergoes a temperature increase ΔT, the bar will experience a thermal strain without any stress

$$\varepsilon_x = \alpha \Delta T$$

The elongation of the bar is

$$\delta = \alpha \Delta T L$$

If the bar is constrained in such a manner that it is not free to expand, the constraints will cause stress to develop.

Example: Consider a bar held between rigid walls that experiences a rise in temperature from 70°F to 200°F as shown in Figure 6-9. Find the stress in the bar.

Figure 6-9

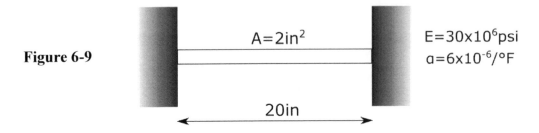

A=2in²   E=30x10⁶psi   a=6x10⁻⁶/°F   20in

We can think of the bar is being subjected to a thermal expansion affect causing an elongation δ_T and a reaction force effect causing a contraction δ_R. Let's consider the thermal expansion effect alone by removing the right support so that the bar is free to expand by the amount δ_T as shown in Figure 6-10.

Figure 6-10

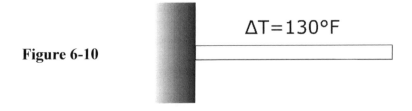

ΔT=130°F

$$\delta_T = \alpha \Delta T L = (6 \times 10^{-6} / °F)(130°F)(20in) = 0.0156in$$

The right wall effectively applies a resisting compressive force R that eliminates the displacement δ_T as shown in Figure 6-11.

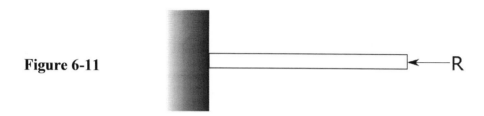

Figure 6-11

The displacement due to the compressive force R is

$$\delta_R = \frac{RL}{EA} = \frac{R(20in)}{(30 \times 10^6 lb / in^2)(2in^2)} = (3.33 \times 10^{-7} in / lb)R$$

Superimposing both the thermal effect and reaction force effect gives the total stretch as

$$\delta_{Total} = \delta_T + \delta_R = 0.0156in + (3.33 \times 10^{-7} in / lb)R = 0$$

Solving for the reaction force gives

$$R = -46,800lb$$

The stress in a bar is solely due to the reaction force R. Therefore,

$$\sigma = \frac{R}{A} = \frac{-46,800lb}{2in^2} = -23,400psi$$

The negative sign indicates compression.

6.5 Torsional Loading of a Bar

Consider a bar with a circular cross-section that has the left end fixed and a torsional moment T at the right end as shown in Figure 6-12.

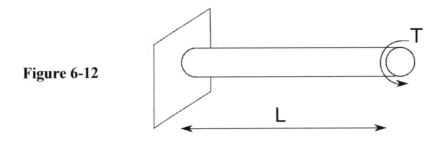

Figure 6-12

A shear stress develops that is uniform along the length but varies linearly from zero at the center to a maximum at the outer radius as shown in Figure 6-13.

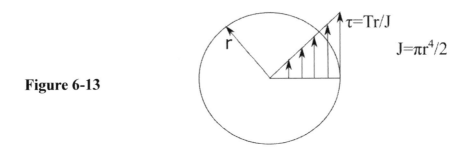

Figure 6-13

The right end will rotate through an angle φ relative to the left end given by

$$\varphi = \frac{TL}{GJ}$$

where G=E/[2(1+υ)] is the shear modulus. These formulas are also valid for a hollow tube with $J = \pi(r_o^4 - r_i^4)/2$.

The logic in handling torsion problems is similar to that for the uniaxial loading case with σ replaced by τ and δ replaced by φ.

6.5.1 Non-uniform bar

Consider a bar with properties G_1 and J_1 in section 1 between points a and b and properties G_2 and J_2 in section 2 between points b and c as shown in Figure 6-14.

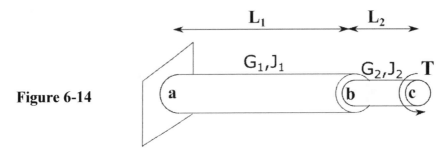

Figure 6-14

We can analyze each section individually. Section 1 experiences an internal torque equal to T which results in

$$\tau_1 = \frac{Tr_1}{J_1} \qquad\qquad \varphi_1 = \frac{TL_1}{G_1 J_1}$$

where τ_1 represents the shear stress in section 1 and φ_1 represents the rotation of point b relative to point a. Section 2 experiences an internal torque equal to T which results in

$$\tau_2 = \frac{Tr_2}{J_2} \qquad\qquad \varphi_2 = \frac{TL_2}{G_2 J_2}$$

where τ_2 represents the shear stress in section 2 and φ_2 represents the rotation of point c relative to point b. The total rotation of the end of the bar (rotation of point c relative to point a) is

$$\varphi = \varphi_1 + \varphi_2 = \frac{TL_1}{G_1 J_1} + \frac{TL_2}{G_2 J_2}$$

If the bar is subjected to multiple loads, we can analyze the bar with each load applied individually and add up the results as was done for bars under uniaxial load.

Example: Find the stress in each section and rotation of the right end of the bar shown in Figure 6-15 if $G=12 \times 10^6$psi.

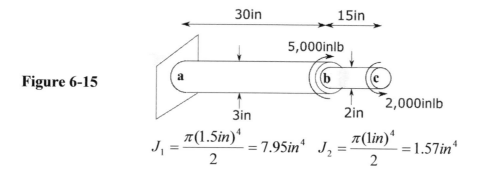

Figure 6-15

$$J_1 = \frac{\pi(1.5in)^4}{2} = 7.95in^4 \quad J_2 = \frac{\pi(1in)^4}{2} = 1.57in^4$$

The effects of each torque will be examined individually. Before beginning, we will settle on a sign convention where counterclockwise is positive and clockwise is negative.

5,000inlb Torque Effect: Consider the 5,000inlb torque acting alone as shown in Figure 6-16.

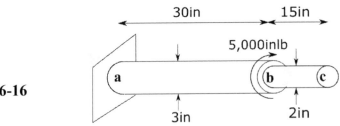

Figure 6-16

Section 1 experiences a 5,000inlb clockwise torque.

$$\tau_1 = \frac{-5,000inlb(1.5in)}{7.95in^4} = -944\,psi$$

$$\varphi_1 = \frac{-5,000\,psi(30in)}{12\times10^6\,lb/in^2(7.95in^4)} = -0.00157rad$$

We recognize that section 2 has zero internal torque which means zero stress and zero rotation.

$$\tau_2 = 0 \qquad \varphi_2 = 0$$

2,000inlb Torque Effect: Consider the 2,000inlb torque acting alone as shown in Figure 6-17.

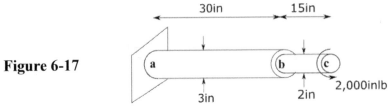

Figure 6-17

Section 1 experiences a 2,000inlb counterclockwise torque resulting in

$$\tau_1 = \frac{2,000inlb(1.5in)}{7.95in^4} = 377\,psi \qquad \varphi_1 = \frac{2,000\,psi(30in)}{12\times10^6\,lb/in^2(7.95in^4)} = 0.000629rad$$

Section 2 also experiences a counterclockwise torque of 2000inlb resulting in

$$\tau_2 = \frac{2,000inlb(1in)}{1.57in^4} = 1,274\,psi \qquad \varphi_2 = \frac{2,000\,psi(15in)}{12\times10^6\,lb/in^2(1.57in^4)} = 0.00159rad$$

Combined Load Effect: We now use the principle of superposition to combine these effects by simply adding the results.

$$\tau_1 = -944\,psi + 377\,psi = -567\,psi$$

$$\tau_2 = 0 + 1,274\,psi = 1,274\,psi$$
$$\varphi_1 = -0.00157\,rad + 0.000629\,rad = -0.000941\,rad$$
$$\varphi_2 = 0 + 0.00159\,rad = 0.00159\,rad$$

The total rotation of point c relative to point a is

$$\varphi_{total} = \varphi_1 + \varphi_2 = -0.000941\,rad + 0.00159\,rad = 0.000649\,rad$$

6.6 Bending of Beams

We can analyze beams using basic principles as we have done in previous sections for bars under uniaxial loading and torsion. Alternatively, we can use tables of known solutions together with the principle of superposition. We will explore the basic principles approach first.

6.6.1 Internal shear force and bending moment in beams

The internal shear force and bending moment are needed to calculate the stresses in the beam. The internal bending moment is also useful for calculating the lateral deflection of the beam. Consider the beam shown in Figure 6-18.

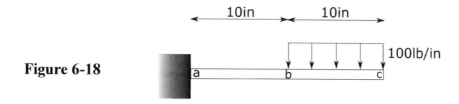

Figure 6-18

We begin by replacing the distributed load by a statically equivalent point load and drawing a free body diagram as shown in Figure 6-19 to find reactions at the support.

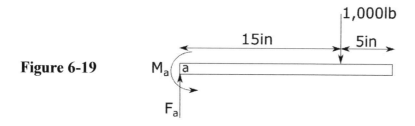

Figure 6-19

Equilibrium requires

$$\sum F_y = 0 \rightarrow F_a - 1000lb = 0$$
$$\sum M_{z-point\,a} = 0 \rightarrow M_a - 1000lb(15in) = 0$$

Solving gives

$$F_a = 1000lb$$
$$M_a = 15,000inlb$$

First, let's find the internal loads between points a and b on the beam by cutting the beam at some point o between these points and drawing a free body diagram of the left-hand piece as shown in Figure 6-20,

Figure 6-20

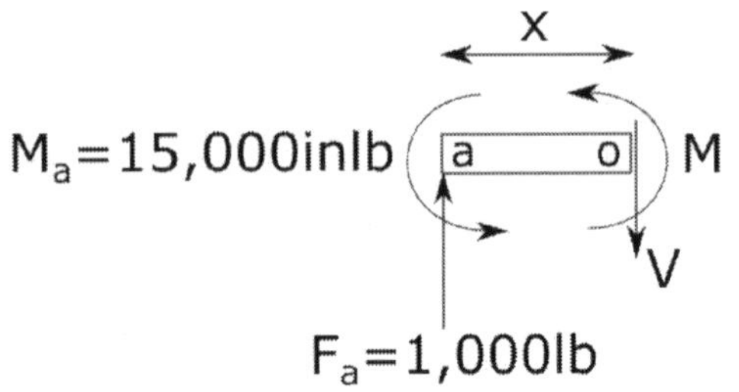

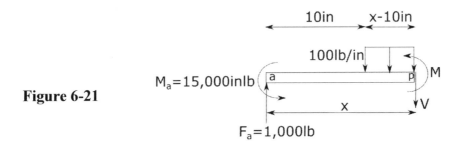

where V is the internal shear force and M is the internal bending moment. For the time being, we will designate the distance between points a and o as x. Equilibrium gives

$$\sum F_y = 0 \rightarrow 1000lb - V = 0$$
$$\sum M_{z-pointo} = 0 \rightarrow 15,000inlb - (1000lb)x + M = 0$$

Solving gives

$$V = 1,000lb$$
$$M = -15,000inlb + (1,000lb)x$$

The free body diagram above remains unchanged for $0 \le x \le 10in$. Therefore, the results for the V and M above are valid formulas for values of x between 0 and 10in. Next let's find the internal loads at some point p between points b and c on the beam by cutting the beam between these points and drawing a free body diagram of the left-hand piece as shown in Figure6-21.

Figure 6-21

Again, we designate the distance between points a and p as x. We now replace the distributed load with a statically equivalent point load as shown in Figure 6-22.

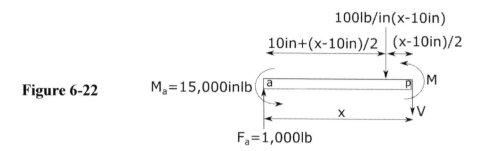

Figure 6-22

Equilibrium requires

$$\sum F_y = 0 \rightarrow 1000lb - 100lb/in(x - 10in) - V = 0$$
$$\sum M_{z-point\,p} = 0 \rightarrow 15,000inlb - (1000lb)x + [100lb/in(x - 10in)](x - 10in)/2 + M = 0$$

Solving gives

$$V = 2,000lb - (100lb/in)x$$
$$M = -15,000inlb + (1,000lb)x - 50lb/in(x - 10in)^2$$

In summary, we of the following formulas for the V and M.

$0 \le x \le 10in$
$$V = 1,000lb \qquad M = -15,000inlb + (1,000lb)x$$

$10 \le x \le 20in$
$$V = 2,000lb - (100lb/in)x \qquad M = -15,000inlb + (1,000lb)x - 50lb/in(x - 10in)^2$$

We can make plots of V and M as functions of position x along the beam as shown in Figure 6-23.

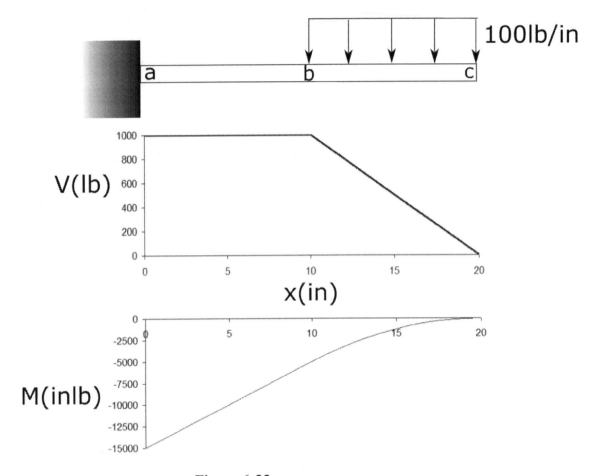

Figure 6-23

The shear force V and moment M are related by

$$\frac{dM}{dx} = V$$

The slope of the curve for M at any point on the beam equals the value of V. The plots above are consistent with this relationship. For $0 \leq x \leq 10in$, V is a constant which means M must be linear with a positive slope in this range. For $10in \leq x \leq 20in$, V is positive but decreases gradually to zero at the end. This means the plot of M must be a curve with a positive slope that gradually becomes flat at the end of the beam.

6.6.2 Stresses in beams

There will be both normal stress σ and shear stress τ in the beam given by the following formulas:

$$\sigma = \frac{My}{I} \quad \text{and} \quad \tau = \frac{VQ}{Ib}$$

The stresses are dependent on the internal loads and the cross-section properties of the beam (in particular, the area moment of inertia I). The area moment of inertia is defined as

$$I = \int_A y^2 dA$$

Fortunately, values of I have been calculated for common cross-sections (see Table 6-2).

Table 6-2 Area Moment of Inertia

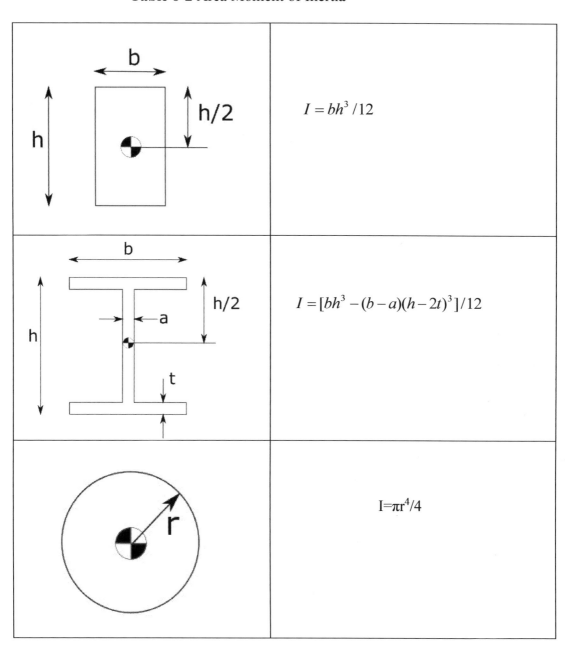

(rectangle with b, h, h/2)	$I = bh^3/12$
(I-beam with b, h, h/2, a, t)	$I = [bh^3 - (b-a)(h-2t)^3]/12$
(circle with r)	$I = \pi r^4/4$

6.6.2.1 Normal stress

To calculate the normal stress, we will use the coordinate x to designate position along the length of the beam and the coordinate y to designate the vertical position on the cross-section *relative to the centroid*. Generally, we are interested in the maximum stress. Therefore, we will select the value of x that makes M a maximum and the point on the cross-section that makes y a maximum.

Example: Find the maximum normal stress for the beam shown in Figure 6-18 if it has the cross-section shown in Figure 6-24.

Figure 6-24

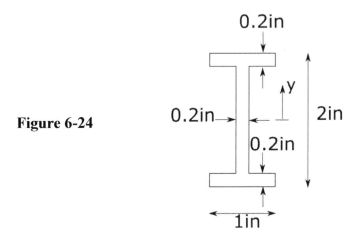

We begin by calculating I for this cross-section.

$$I = [1in(2in)^3 - (1in - 0.2in)(2in - 2(0.2in))^3]/12 = 0.3936in^4$$

The maximum value of M is 15,000inlb which occurs at the support (x=0). The maximum value of y is 1in which occurs at the top of the cross-section. Therefore,

$$\sigma = \frac{My}{I} = \frac{15,000inlb(1in)}{0.3936n^4} = 38,110psi$$

6.6.2.2 Shear stress

The maximum value of shear stress occurs where the value of x that makes V a maximum and the location on the cross-section that makes Q a maximum (this generally occurs at the centroid). The most straightforward way to calculate Q is the following:
1. Draw a horizontal line through the cross-section at the point where τ is desired (in this case at the centroid).
2. Select either the area above or below this line and divide it into one or more rectangular pieces.

3. For each rectangular piece, form the product of the area A_i of the piece times the vertical distance $\bar{y}_i$ from the cross-section centroid to the center of the rectangular piece.
4. Add up the results from step 3 to find Q

$$Q = \sum_{i=1}^{n} A_i \bar{y}_i$$

<u>Example</u>: Find the maximum shear stress for the same beam used in the previous example to calculate normal stress.

The maximum value of V is 1,000lb and occurs at the support (x=0). To calculate Q for this case, we draw a horizontal line through the centroid and take the two rectangular areas above it as shown in Figure 6-25.

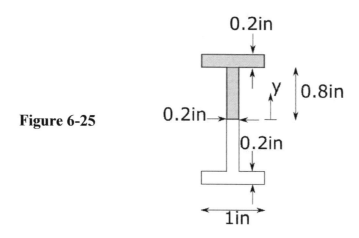

Figure 6-25

We set

$$Q = A_1\bar{y}_1 + A_2\bar{y}_2 = [(0.2in)(0.8in)](0.4in) + [(1in)(0.2in)](0.9in) = 0.244in^3$$

Therefore, the shear stress is

$$\tau = \frac{VQ}{Ib} = \frac{1,000lb(0.244in^3)}{0.3936in^4(0.2in)} = 3,100\,psi$$

6.6.3 Beam deflections

To calculate the lateral deflection of a beam, we need a formula for the internal bending moment M as a function of position x. The deflection v is related to M through

$$\frac{d^2v}{dx^2} = \frac{M}{EI}$$

This relationship must be integrated twice and boundary conditions applied to arrive at a formula for v as a function of position x.

Example: Find the deflection of the beam shown in Figure 6-26.

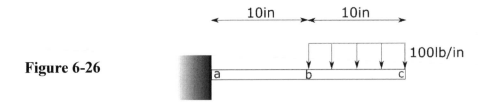

Figure 6-26

In the previous section, we arrived at the following formulas for M.

$$0 \leq x \leq 10in \quad M = -15,000inlb + (1,000lb)x$$
$$10 \leq x \leq 20in \quad M = -15,000inlb + (1,000lb)x - 50lb/in(x-10in)^2$$

Since we have different formulas for different sections of the beam, we have to analyze the two sections separately.

Section 1: $0 \leq x \leq 10in$

The deflection formula gives

$$\frac{d^2v}{dx^2} = \frac{M}{EI} \rightarrow \frac{d^2v}{dx^2} = \frac{-15,000inlb + (1,000lb)x}{EI}$$

We begin by performing an indefinite integration which involves adding a constant of integration to give

$$\frac{dv}{dx} = \frac{(-15,000inlb)x + (1,000lb)x^2/2}{EI} + C_1$$

To find C_1, we apply the boundary condition that the slope of the beam must be zero at the fixed end; i.e.,

$$\frac{dv}{dx}\bigg|_{x=0} = -\frac{(-15,000inlb)(0) + (1,000lb)(0)^2/2}{EI} + C_1 = 0 \rightarrow C_1 = 0$$

We conclude that the constant C_1 must be zero. Next, we integrate again adding a constant of integration to get

$$v = \frac{(-15,000inlb)x^2/2 + (1,000lb)x^3/6}{EI} + C_2$$

To find C_2, we must apply the boundary condition that the deflection of the beam at the fixed end must be zero; i.e.,

$$v|_{x=0} = -\frac{(-15,000inlb)(0)^2/2 + (1,000lb)(0)^3/6}{EI} + C_2 = 0 \rightarrow C_2 = 0$$

We conclude that the constant C_2 must be zero. To find the deflection in the range $0 \leq x \leq 10in$, we simply substitute the desired value of x into

$$v = \frac{(-15,000inlb)x^2/2 + (1,000lb)x^3/6}{EI}$$

Section 2: $10 \leq x \leq 20in$

To find a similar formula for that is valid in this section, we would need to repeat the same operations on

$$\frac{d^2v}{dx^2} = \frac{-15,000inlb + (1,000lb)x + 50lb/in(x-10in)^2}{EI}$$

6.6.4 Solving problems using tables

A considerable amount of effort can be saved by using tables of known solutions (see Table 6-3) to beam problems coupled with the principle of superposition.

Example: Find the deflection of the right end and the maximum normal and shear stress at the left end of the beam shown in Figure 6-27.

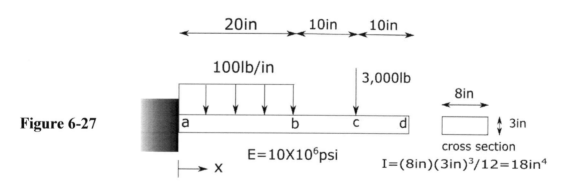

Figure 6-27

Consider the beam with the 100lb/in uniformly distributed load acting alone as shown in Figure 6-28.

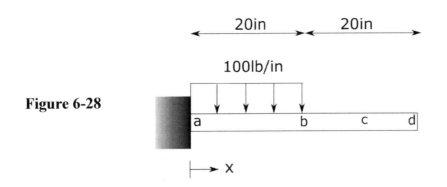

Figure 6-28

Table 6-3 Beam Deflection

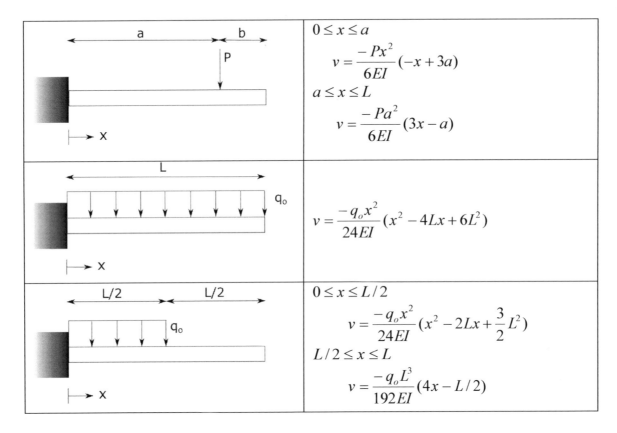

From the table, the displacement in the range $0 \le x \le L/2$ is given as

$$v = \frac{-q_o x^2}{24EI}(x^2 - 2Lx + \tfrac{3}{2}L^2)$$

We get the moment by differentiating this twice

$$M = EI\frac{d^2 v}{dx^2} = \frac{-q_o}{24}\frac{d^2}{dx^2}(x^4 - 2Lx^3 + \tfrac{3}{2}L^2 x^2)$$

$$M = \frac{-q_o}{24}\frac{d}{dx}(4x^3 - 6Lx^2 + 3L^2 x)$$

$$M = \frac{-q_o}{24}(12x^2 - 12Lx + 3L^2)$$

We get the shear force by differentiating this once

$$V = \frac{dM}{dx} = \frac{-q_o}{24}(24x - 12L)$$

At the support (x=0), $M = \frac{-100lb/in}{24}[3(40in)^2] = -20,000inlb$

$$V = \frac{-100lb/in}{24}[-12(40in)] = 2,000lb$$

We write the displacement for the next section $L/2 \leq x \leq L$

$$v = \frac{-q_o L^3}{192EI}(4x - L/2)$$

At the right end (x=40in), $v = \frac{-100lb/in(40in)^3}{192(10\times10^6 lb/in^2)(18in^4)}[4(40in) - 40in/2)] = -0.0259in$

Consider the beam with the 3,000lb concentrated load acting alone as shown in Figure 6-29.

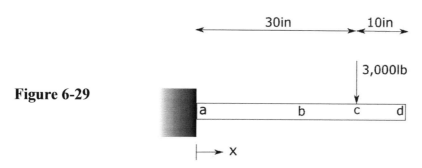

Figure 6-29

From the table we have the displacement in the range $0\leq x \leq a$ given as

$$v = \frac{-Px^2}{6EI}(-x + 3a)$$

As before, we get the moment by differentiating twice.

$$M = EI\frac{d^2v}{dx^2} = \frac{-P}{6}\frac{d^2}{dx^2}(-x^3 + 3ax^2)$$

$$M = \frac{-P}{6}\frac{d}{dx}(-3x^2 + 6ax) = \frac{-P}{6}(-6x + 6a)$$

We get the shear force by differentiating the moment.

$$V = \frac{dM}{dx} = P$$

At the support (x=0), $M = \frac{-3,000lb}{6}[6(30in)] = -90,000inlb$

$$V = 3,000lb$$

We write the displacement for the next section $a \le x \le L$

$$v = \frac{-Pa^2}{6EI}(3x - a)$$

At the right end (x=40in)

$$v = \frac{-3000lb(30in)^2}{6(10 \times 10^6\ psi)(18in^4)}[3(40in) - 30in)] = -0.225in$$

Using superposition, the total deflection of the right end is

$$v = -0.0259 - 0.225in = -0.251in$$

At the left end

$$M = -20,000inlb - 90,000inlb = -110,000inlb$$
$$V = 2,000lb + 3,000lb = 5,000lb$$

From these we get the stresses as

Normal stress at y=1.5in $\sigma = \frac{My}{I} = \frac{110,000inlb(1.5in)}{18in^4} = 9,167\,psi$

Shear stress at y=0 $\tau = \frac{VQ}{Ib} = \frac{5,000lb(9in^3)}{18in^4(8in)} = 313\,psi$

6.7 Buckling of Bars

Slender bars under uniaxial compressive load are susceptible to failure by buckling, which is an unstable response, resulting in a lateral deflection. The critical load to cause buckling depends on the way that the bar is supported. For a beam with simple (hinged) supports at both ends as shown in Figure 6-30, the critical load is $P_{cr} = \pi^2 EI/L^2$.

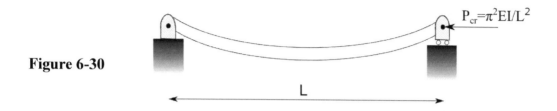

Figure 6-30

We can use the same formula for other support conditions if we replace the actual length L by an effective length KL; i.e.,

$$P_{cr} = \frac{\pi^2 EI}{(KL)^2}$$

The values of K for the various cases are as follows:
1) Both ends hinged: K=1
2) Both ends fixed: K=0.5
3) One end fixed and the other end hinged: K=0.7
4) One end fixed and the other end free: K=2

Example: A bar with a circular cross-section and E=30x10^{6}psi must carry a 10,000lb uniaxial load as shown in Figure 6-31. Find the minimum diameter d to prevent buckling.

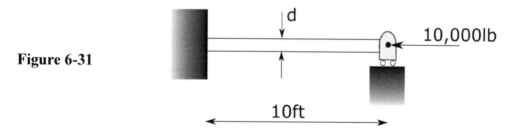

Figure 6-31

For this case, K=0.7. Let's rewrite the critical load equation as follows:

$$I = \frac{P_{cr}(0.7L)^2}{\pi^2 E}$$

$$\frac{\pi r^4}{4} = \frac{P_{cr}(0.7L)^2}{\pi^2 E}$$

$$r = \left(\frac{4P_{cr}(0.7L)^2}{\pi^3 E}\right)^{1/4} = \left(\frac{4(10,000lb)[0.7(120in)]^2}{\pi^3(30\times10^6)}\right)^{1/4} = 0.743in$$

7. DYNAMICS

Dynamics deals with the study of bodies in motion. These bodies can be divided into two categories: particles and rigid bodies. Bodies whose size and shape are not important factors in the analysis are classified as particles. For example, the size and shape of the car are irrelevant in calculating the travel time between two cities. Bodies whose size and shape are important factors are classified as rigid bodies. For example, the size and shape of the car are very relevant when parking the car. Furthermore, we can classify dynamics problems as either kinematics (motion is studied without concern for the forces causing the motion) and kinetics (where the forces are accounted for).

7.1 Kinematics of Particles

7.1.1 Rectangular coordinates

The position of a particle can be specified relative to a stationary reference frame by specifying x,y and z-coordinates of the particle. These coordinates will vary with time as the particle moves through space. The velocity of the particle will have x, y, and z components that depend on the time derivatives of the coordinates.

$$v_x = \frac{dx}{dt} = \dot{x} \qquad v_y = \frac{dy}{dt} = \dot{y} \qquad v_z = \frac{dz}{dt} = \dot{z}$$

These components form a velocity vector whose magnitude is

$$v = \sqrt{v_x^2 + v_y^2 + v_z^2}$$

The rate at which velocity components change provides the components of the acceleration of the particle as

$$a_x = \frac{dv_x}{dt} = \dot{v}_x \qquad a_y = \frac{dv_y}{dt} = \dot{v}_y \qquad a_z = \frac{dv_z}{dt} = \dot{v}_z$$

Example: Find the magnitude of the acceleration of a particle moving in the x,y-plane whose coordinates are given by x=(10ft/s²) t² and y=(2ft/s³)t³ for t=3s.

Differentiating once with respect to t gives the velocity components as

$$v_x = 2(10\,ft/s^2)t \qquad v_y = 3(2\,ft/s^3)t^2$$

Differentiating a second time gives the acceleration components as

$$a_x = 20\,ft/s^2 \qquad a_y = 6(2\,ft/s^3)t$$

Thus,

$$a_y = 12\,ft\,/\,s^3(3s) = 36\,ft\,/\,s^2$$
$$a = \sqrt{a_x^2 + a_y^2} = \sqrt{(20\,ft\,/\,s^2)^2 + (36\,ft\,/\,s^2)^2} = 41.2\,ft\,/\,s^2$$

7.1.1.1 Straight-line motion

For a particle moving in a straight line, we can orient the x-axis along this line so that the position, velocity, and acceleration have x-components only. For the special case where the acceleration is a constant a, the initial position is x_0, and the initial velocity is v_0, we have the following relation for position and velocity as the particle moves:

$$x = x_0 + v_0 t + \frac{1}{2}at^2$$

$$v = v_0 + at \quad \text{or} \quad v = \sqrt{v_0^2 + 2a(x - x_0)}$$

Example: If the brakes on a car are capable of causing a deceleration of 30ft/s², find the distance required to stop the car from a velocity of 60mph.

We have v_0=(60mile/hr)(5,280ft/mile)(1hr/3600s)=88ft/s and final velocity v=0. Then

$$0 = \sqrt{v_0^2 + 2ax}$$

$$x = -\frac{v_0^2}{2a} = -\frac{(88\,ft\,/\,s)^2}{2(-30\,ft\,/\,s^2)} = 129\,ft$$

7.1.1.2 Projectile motion

An object that is launched at an angle θ and initial velocity v_0 will experience a downward acceleration of g as it flies through the air as shown in Figure 7-1.

Figure 7-1

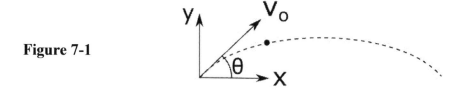

Velocity and position at time t are given as

$$x = (v_o \cos \theta)t \qquad\qquad y = (v_o \sin \theta)t - \frac{1}{2}gt^2$$

$$v_x = v_o \cos \theta \qquad\qquad v_y = v_o \sin \theta - gt$$

Example: The outfielder on a baseball field wants to throw a ball to home plate which is 250ft away. If he launches the ball with initial velocity of v_0=110ft/s, find the necessary launch angle θ for the ball to reach the plate.

Solving for t from the equations above gives

$$t = \frac{x}{v_o \cos \theta}$$

We use the fact that the finishing value of y is zero.

$$0 = (v_o \sin \theta)t - \frac{1}{2}gt^2$$

Then

$$\frac{g}{2}\left(\frac{x}{v_o \cos \theta}\right) = v_o \sin \theta$$

$$\frac{g}{2}x = v_o^2 \sin \theta \cos \theta$$

We use the trig identity

$$\sin \theta \cos \theta = \frac{1}{2}\sin 2\theta$$

to get

$$\sin 2\theta = \frac{g}{v_o^2}x = \frac{32.2\,ft/s^2}{(110\,ft/s)^2}(250\,ft) = 0.665$$

$$\theta = 20.9°$$

7.1.2 Polar coordinate components

For certain problems, especially those involving objects moving in circular arcs, it is convenient to describe the motion in terms of a radial distance r and an angular position θ as shown in Figure 7-2.

Figure 7-2

For this case, the components of velocity and acceleration are as follows:

$$v_r = \dot{r} \qquad a_r = \ddot{r} - r\dot{\theta}^2$$
$$v_\theta = r\dot{\theta} \qquad a_\theta = r\ddot{\theta} + 2\dot{r}\dot{\theta}$$

<u>Example</u>: A disc rotates in a vertical plane as shown in Figure 7-3. A block connected to a string that remains taut is free to slide in the slot. The distance between the block and the center of the disc is 2ft. Find the magnitude of the acceleration experienced by the block if $\dot{\theta} = 5 rad / s$ and $\ddot{\theta} = 20 rad / s$.

Figure 7-3

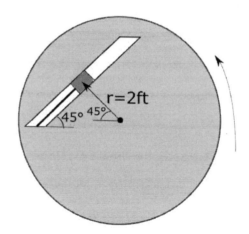

Since the radial distance remains constant as the disc rotates, we conclude

$$\dot{r} = 0 \qquad \ddot{r} = 0$$

Therefore,

$$a_r = -r\dot{\theta}^2 = -2 ft(5s^{-1})^2 = -50 ft / s^2$$
$$a_\theta = r\ddot{\theta} = 2 ft(20s^{-2}) = 40 ft / s^2$$
$$a = \sqrt{a_r^2 + a_\theta^2} == \sqrt{(-50 ft / s^2)^2 + (40 ft / s^2)^2} = 64 ft / s^2$$

7.1.3 Normal and tangential components

An object traveling along an arc with radius of curvature ρ will have a velocity v tangent to the arc. There will be two components of acceleration: a component tangent to the arc a_t and a component normal to the arc a_n, as shown in Figure 7-4.

Figure 7-4

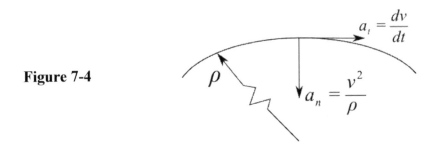

Example: A car traveling at 60mph (88ft/s) passes over the crest of a hill with a radius of curvature of 500ft as shown in Figure 7-5. The driver applies the brakes slowing the car at a rate of 20ft/s². Find the magnitude of the acceleration of the car.

Figure 7-5

We have the following relations

$$a_t = \dot{v} = -20\, ft/s^2$$

$$a_n = \frac{v^2}{\rho} = \frac{(88\, ft/s)^2}{500\, ft} = 15.5\, ft/s^2$$

The resultant acceleration is

$$a = \sqrt{a_t^2 + a_n^2} = 25.3\, ft/s^2$$

7.2 Particle Kinetics

Particle kinetics problems are solved using Newton's second law of motion, which states that the net force on an object equals the mass times the acceleration.

$$\mathbf{F} = m\mathbf{a}$$

We have different versions of this in component form depending on the coordinate system chosen:

$$\sum F_x = ma_x \text{ and } \sum F_y = ma_y$$

or $\quad \sum F_r = ma_r \text{ and } \sum F_\theta = ma_\theta$

or $\quad \sum F_t = ma_t \text{ and } \sum F_n = ma_n$

The forces that appear on the left side of the equations are determined from a free body diagram similar to what was done for statics problems. The problems are readily solved through the following solution steps:

1. Draw a free body diagram

2. Select a coordinate system that is convenient for writing the acceleration components
3. Use the free body diagram to express the force components on the left side of the equations
4. Use the appropriate acceleration formula to express the acceleration components on the right side of the equations
5. Apply Newton's second law equations above
6. Solve for unknown quantities

7.2.1 Rectangular coordinate components

Example: A force pushes a block weighing 5 pounds up an inclined plane with a coefficient of friction of $\mu=0.3$ as shown in Figure 7-6. Find the acceleration of the block.

Figure 7-6

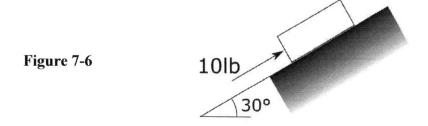

10lb

30°

The mass of the block is m=5/32.2slugs=0.155slugs.

We begin by drawing a free body diagram of the block as shown in Figure 7-7.

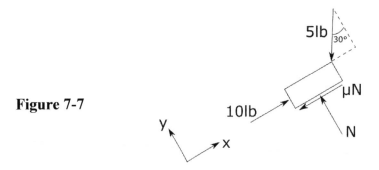

Figure 7-7

We align the x,y-coordinate system with the inclined plane and apply Newton's law.

$$\sum F_x = ma_x \rightarrow 10lb - (5lb)\sin 30° - 0.3N = (0.155slug)a_x$$
$$\sum F_y = ma_y \rightarrow N - (5lb)\cos 30° = 0$$

Solving gives

$$N = 4.33lb \text{ and } a_x = 40 ft/s^2$$

7.2.2 Polar coordinate components

Example: If the block in Figure 7-3 weighs 2lb and the surfaces are frictionless, find the tension in the string.

We begin with a free body diagram of the block as shown in Figure 7-8.

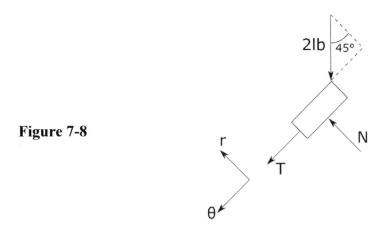

Figure 7-8

The mass of the block is m=2/32.2slug=0.062slug. Recall that a_θ=40ft/s². Newton's law gives

$$\sum F_\theta = ma_\theta \rightarrow T + (2lb)\sin 45° = (0.062slug)(40 ft/s^2)$$

Solving gives

$$T = 1.07 lb$$

7.2.3 Normal and tangential components

Example: A car travels over the crest of a hill which has a radius of curvature of 200 feet as shown in Figure 7-9. Find the maximum speed that the car can travel without becoming airborne.

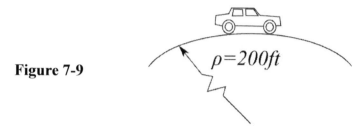

Figure 7-9

We begin by drawing a free body diagram of the car as shown in Figure 7-10.

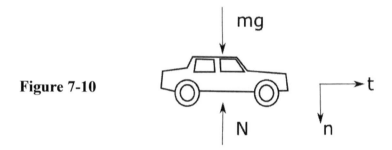

Figure 7-10

Newton's law gives

$$\sum F_n = ma_n \rightarrow -N + mg = m\frac{v^2}{\rho}$$

When the car is about to become airborne, N equals zero. When this occurs, the velocity is

$$v = \sqrt{\rho g} = \sqrt{200\, ft(32.2\, ft\,/\,s^2)} = 80.2\, ft\,/\,s$$

7.2.4 Work and energy

For certain kinetics problems, the analysis can be simplified by using an integrated form of Newton's second law. For problems involving velocity and distance, a work-energy approach works well. The work done by a force equals the force integrated along the path traveled from the starting point to the finishing point;

$$Work = \int \vec{F} \cdot d\vec{r}$$

If the force is constant, this simplifies to

$$Work = Fd \cos\theta$$

where F is the magnitude of the force, d is the distance from the starting point to the finishing point, and θ is the angle between F and d. If F is in the same direction as d, then work=Fd. If F is opposite to the direction of motion (for example, a friction force), work=-Fd.

The work done on a particle equals the change in kinetic energy of the particle.

$$TotalWork = \frac{1}{2}mv_f^2 - \frac{1}{2}mv_s^2$$

where v_s is the starting velocity, and v_f is the finishing velocity. The work done by the gravity force can be accounted for by using a change in potential energy given by

$$P.E. = mgh$$

where h is the elevation of the particle. The work done by a spring force can be accounted for by using a change in potential energy given by

$$P.E. = \frac{1}{2}k\delta^2$$

where k is the spring stiffness and δ is the stretch of the spring. The relationship for the general case becomes

$$d\sum_{i=1}^{n}(F_i \cos\theta_i) = \frac{1}{2}m(v_f^2 - v_s^2) + \frac{1}{2}k(\delta_f^2 - \delta_s^2) + mg(h_f - h_s)$$

where the F_i are the constant forces, and the subscript f refers to finishing value, and the subscript s refers to starting value.

Example: A block weighing 20 pounds starts from rest and slides down the incline with a coefficient of friction of μ=0.2 as shown in Figure 7-11. Find the velocity after it has traveled 10ft.

81

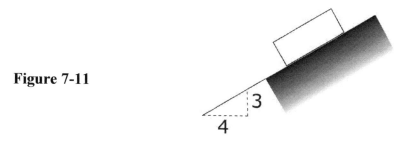

Figure 7-11

We begin by drawing a free body diagram of the block.

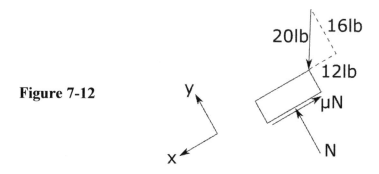

Figure 7-12

The block does not move in the y-direction. Therefore, the normal force N balances the 16lb component of gravity. The friction force is

$$f = \mu N = 0.2(16lb) = 3.2lb$$

The only forces doing work are those in the x-direction; i.e., the 12lb component of the gravity force and the friction force.

$$dF = \frac{1}{2}mv_f^2 \quad \rightarrow \quad 10\,ft(12lb - 3.2lb) = \frac{1}{2}\left(\frac{20}{32.2}\,slug\right)v_f^2$$

Solving gives

$$v_f = 16.8\,ft/s$$

We could have accounted for the gravity force using mgh=(20lb)(6ft)=120ftlb.

<u>Example</u>: A 2kg collar slides along a 1m diameter circular hoop in a vertical plane under the action of a 20Newton constant horizontal force as shown in Figure 7-13. The unstretched length of the spring is 0.5m, and it has a stiffness of 200Newton/m. If the collar starts from rest at position a, find the velocity at position b.

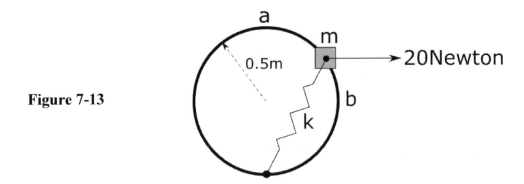

Figure 7-13

The starting elevation of the collar is 1m. The final elevation is 0.5m. The starting length of the spring is 1m. The finishing length of the spring is 0.707m. The distance d between the starting point and the finishing point is 0.707m. The angle between the 20Newton force and d is 45°. The work-energy equation becomes

$$dF\cos\theta = \frac{1}{2}m(v_f^2 - v_s^2) + \frac{1}{2}k(\delta_f^2 - \delta_s^2) + mg(h_f - h_s)$$

$$0.707m(20Newton)\cos 45° = \frac{1}{2}(2kg)v_f^2 + \frac{1}{2}(200Newton/m)[(0.707m - 0.5m)^2 - (1m - 0.5m)^2]$$

$$+ 2kg(9.81m/s^2)(0.5m - 1m)$$

Solving gives

$$v_f = 6.37/s$$

7.2.5 Impulse and momentum

For kinetics problems involving time and velocity, impulse and momentum concepts can simplify the analysis. The impulse of a force equals the integral of the force over time.

$$impulse = \int_{t_s}^{t_f} \mathbf{F}dt$$

The momentum is defined as the mass times velocity

$$momentum = m\mathbf{v}$$

The impulse equals the change in momentum;

$$\int_{t_s}^{t_f} F_x dt = mv_{xf} - mv_{xs}$$

$$\int_{t_s}^{t_f} F_y \, dt = mv_{yf} - mv_{ys}$$

where F_x and F_y are the x and y-components of force, respectively; and v_{fx} and v_{fy} are the x and y-components of the finishing velocity, respectively; and v_{sx} and v_{sy} are the x and y-components of the starting velocity, respectively.

Example: A constant 50Newton force acts on a 5kg block that slides along a horizontal surface with coefficient of friction of µ=0.3 as shown in Figure 7-14. If the block has an initial velocity of 10m/s, find its velocity after 5s.

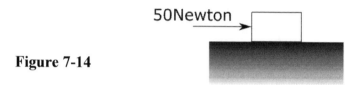

Figure 7-14

We begin by drawing a free body diagram as shown in Figure 7-15.

Figure 7-15

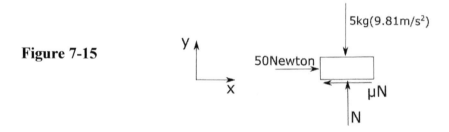

Since the block does not move in the vertical direction, the normal force N simply balances the gravitational force mg=5kg(9.81m/s²)=49.1Newton. Thus, the friction force is µN=0.3(49.1Newton)=14.7Newton. Applying the impulse-momentum relation for the x-direction, we get

$$\int F_x \, dt = mv_{xf} - mv_{xs}$$

$$\int_0^{5s} (50 Newton - 14.7 Newton) dt = (5kg)v_{xf} - (5kg)(10m/s)$$

Solving for v_{fx} gives

$$v_{xf} = 10m/s + (50 Newton - 14.7 Newton)(5s)/5kg = 45.3m/s$$

7.2.5.1 Conservation of momentum

For the case of a system of particles that do not experience external forces (i.e., the only forces are interactive forces between particles), the total momentum of the system must be conserved; i.e.,

$$\sum_{i=1}^{n} m_i v_{ixf} = \sum_{i=1}^{n} m_i v_{ixs}$$

$$\sum_{i=1}^{n} m_i v_{iyf} = \sum_{i=1}^{n} m_i v_{iys}$$

Example: An astronaut with a mass of 100kg carries a pack with a mass of 20kg. If he is initially at rest, find his velocity after he pushes the pack away with the velocity of 10m/s.

We recognize that the only forces are internal. Since the starting momentum is zero, the finishing momentum must also be zero. Thus, we conclude

$$m_{astro} v_{astro} + m_{pack} v_{pack} = 0$$

$$v_{astro} = -\frac{m_{pack}}{m_{astro}} v_{pack} = -\frac{20kg}{100kg}(10m/s) = -2m/s$$

7.2.5.2 Collisions

Consider the case of two particles colliding as shown in Figure 7-16.

Figure 7-16

Before Collision   After Collision

Conservation of momentum requires

$$m_1 v_{xf1} + m_2 v_{xf2} = m_1 v_{xs1} + m_2 v_{xs2}$$

Kinetic energy is generally not conserved. This is quantified through the coefficient of restitution e, where e=1 corresponds to a perfectly elastic collision, and e=0 corresponds to the particles sticking together. The velocities of the two particles after the collision are related to the velocities before the collision according to

$$v_{1xf} = \frac{(1+e)m_2 v_{2xs} + (m_1 - em_2)v_{1xs}}{m_1 + m_2}$$

$$v_{2xf} = \frac{(1+e)m_1 v_{1xs} + (m_2 - em_1)v_{2xs}}{m_1 + m_2}$$

Example: The head of a golf club with a mass of 190grams has a velocity of 50m/s. It strikes a stationary ball with a mass of 46grams. Find the velocity of the ball after impact if e=0.8.

For this case we have

$$v_{ball} = \frac{(1+e)m_{club}v_{club}}{m_{ball} + m_{club}} = \frac{(1+0.8)(190grams)(50m/s)}{46grams + 190grams} = 72.5m/s$$

7.3 Rigid Body Kinematics

Consider the motion of a rigid body in the form of the rectangular plate as it moves from position 1 to position 2 as shown in Figure 7-17.

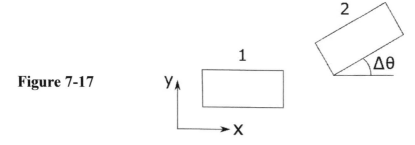

Figure 7-17

To fully describe this motion, we need to specify both a translation and a rotation. We have already dealt with translation in studying particle motion. We now need to account for the rotation. In particular, we are interested in the rate of rotation; i.e., the angular velocity ω.

$$\omega = \frac{\Delta\theta}{\Delta t} \rightarrow \omega = \frac{d\theta}{dt} = \dot\theta \qquad \text{the units for } \omega \text{ are radians per second}$$

where we designate counterclockwise rotation as positive and clockwise rotation as negative.

7.3.1 Velocity analysis using components

When a body is rotating, each point on the body will have a different translational velocity v. If we consider two points p and q on the body and establish a coordinate system at point q as shown in Figure 7-18, the velocity components of these two points are related as follows:

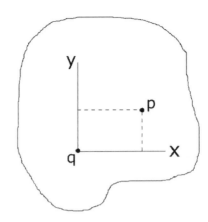

Figure 7-18

$$v_{px} = v_{qx} - \omega y$$
$$v_{py} = v_{qy} + \omega x$$

where x and y are the coordinates of point p relative to point q. These relations form the basis for kinematic analysis of rigid bodies.

Example: A ladder with length L=8ft leans against a vertical wall as shown in Figure 7-19. If the bottom of the ladder slides to the right at a constant velocity of 3ft/s, find the velocity of the top of the ladder and the angular velocity of the ladder when θ=60°.

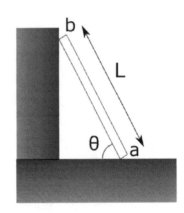

Figure 7-19

Let's put point q at the bottom of the ladder (at point a) and point p at the top (at point b). The coordinates of point b are

$$x = -L\cos\theta = -8\,ft\cos 60° = -4\,ft \quad y = L\sin\theta = 8\,ft\sin 60° = 6.93\,ft$$

We recognize that point b has vertical motion only (v_{bx}=0), and point a has horizontal motion only (v_{ay}=0). The velocity equations are

$$v_{bx} = v_{ax} - \omega y \to 0 = 3\,ft/s - \omega(6.93\,ft)$$
$$v_{by} = v_{ay} + \omega x \to v_{by} = 0 + \omega(-4\,ft)$$

Solving gives

$$\omega = 0.433s^{-1} \quad v_{by} = -1.73\,ft/s$$

We can use these results to calculate the velocity of any point on the ladder.

7.3.2 Velocity analysis using the instant center of rotation

For a body that is pinned at a point o and rotates about that point, the analysis is particularly simple. Every point in the body moves in a circular path about the point o with radius r_{oa} (which is the distance between the point o and the point a as shown in Figure 7-20).

Figure 7-20

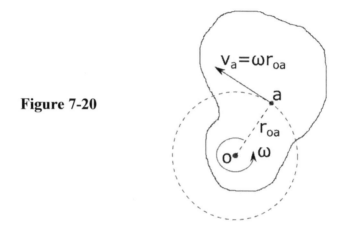

From our study of circular motion, we know that the velocity of point a is $v_a=\omega r_{oa}$ and is directed perpendicular to the radial line r_{oa}. We can apply these ideas to the general case (where the body is not pinned at a point) if we imagine that the rotation point o exists for an instant in time (i.e., it exists only for a specific position and changes its location as the body changes position). The velocity of any point on the body will be perpendicular to the radial line from the rotation point o to the moving point a. We can locate this instant center of rotation if we know the direction of the velocity of two different points on the body. We simply draw lines perpendicular to the velocity vectors at these two points. The instant center will be located at the intersection of these two perpendicular lines as shown in Figure 7-21.

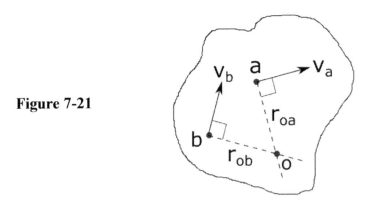

Figure 7-21

We will then use geometry principles to determine the radial distances r_{oa} and r_{ob}. Then, we have the relations

$$v_a = \omega r_{oa}$$
$$v_b = \omega r_{ob}$$

We need to keep in mind that the instant center point does not need to be on the body being analyzed. It could be a point in empty space or even a point on another body. For example, consider a bar with wheels on the ends that travel around a circular slot as shown in Figure 7-22.

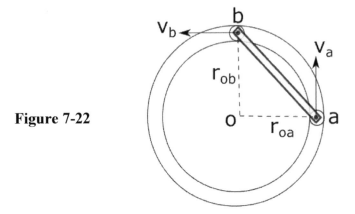

Figure 7-22

It is clear that the velocity of point a is vertical and the velocity of point b is horizontal. The lines perpendicular to these two velocities intersect at point o, indicating that this point is the instant center for the bar. However, even without the knowledge of instant center principles, it is probably clear to the reader that as the wheel travels around the slot, the bar rotates about point o. We might even pretend that there is an imaginary bar pinned at point o that extends up to the center of bar ab where it is welded. In this case, the location of the instant center does not change as the bar moves. In general, we will see that the location of the instant center for a body will move as the body moves.

Example: We will apply this technique to the sliding ladder in Figure 7-19. We know that the velocity of the bottom (point a) must be horizontal and the velocity of the top (point b) must be vertical. We draw perpendicular lines to the two velocity vectors as shown in Figure 7-23 to locate the instant center at point o.

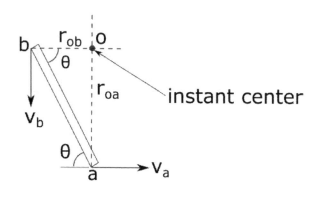

Figure 7-23

For θ=60°, we have

$$r_{oa} = 8\,ft\sin 60° = 6.93\,ft \qquad r_{ob} = 8\,ft\cos 60° = 4\,ft$$

We are given v_a=3ft/s . Thus,

$$\omega = \frac{v_a}{r_{oa}} = \frac{3\,ft/s}{6.93\,ft} = 0.433s^{-1}$$

$$v_b = \omega r_{ob} = 0.433s^{-1}(4\,ft) = 1.73\,ft/s$$

The distances r_{oa} and r_{ob} depend on θ. Thus, the location of the instant center would change as the bar moves.

7.3.2.1 Four-bar linkage

Example: For the four-bar linkage shown in Figure 7-24, the angular velocity of bar ab is 10rad/s counterclockwise. Find the angular velocities of bars bc and cd.

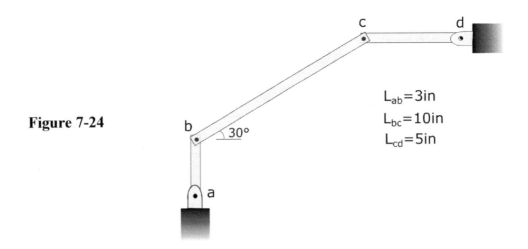

Figure 7-24

$L_{ab}=3in$
$L_{bc}=10in$
$L_{cd}=5in$

Because bar ab rotates about the pin at a, the velocity of point b must be horizontal as shown in Figure 7-25.

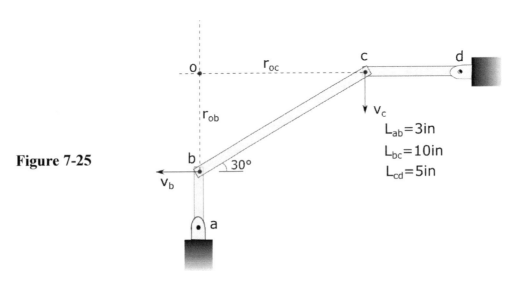

Figure 7-25

$L_{ab}=3in$
$L_{bc}=10in$
$L_{cd}=5in$

Because bar cd rotates about the pin at d, the velocity of point c must be vertical. For bar bc, we know the direction of the velocity of two points (b and c) on the bar. Constructing lines perpendicular to v_b and v_c gives the instant center for bar bc at point o. From simple trigonometry we can calculate the following distances:

$$r_{ob} = 10in \sin 30° = 5in \qquad r_{oc} = 10in \cos 30° = 8.66in$$

We now perform the following series of calculations.

$$v_b = \omega_{ab}r_{ab} = 10s^{-1}(3in) = 30in/s$$
$$\omega_{bc} = \frac{v_b}{r_{ob}} = \frac{30in/s}{5in} = 6s^{-1}$$

$$v_c = \omega_{bc}r_{oc} = 6s^{-1}(8.66in) = 51.96in/s$$

$$\omega_{cd} = \frac{v_c}{r_{cd}} = \frac{51.96in/s}{5in} = 10.4s^{-1}$$

Some general relations have been developed for four-bar linkage mechanisms. With angles θ_2, θ_3, and θ_4 defined as shown in Figure 7-26, we have the following formulas.

Figure 7-26

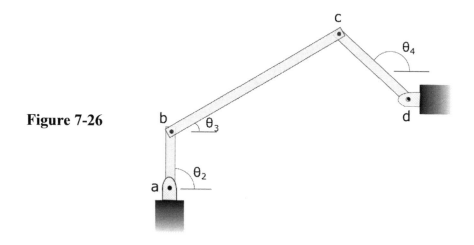

$$\omega_{bc} = \omega_{ab}\frac{r_{ab}}{r_{bc}}\frac{\sin(\theta_4 - \theta_2)}{\sin(\theta_3 - \theta_4)}$$

$$\omega_{cd} = \omega_{ab}\frac{r_{ab}}{r_{cd}}\frac{\sin(\theta_2 - \theta_3)}{\sin(\theta_4 - \theta_3)}$$

For our example, we have θ_2=90°, θ_3=30°, and θ_4=180°. The formulas give

$$\omega_{bc} = (10s^{-1})\frac{3in}{10in}\frac{\sin(180° - 90°)}{\sin(30° - 180°)} = -6s^{-1}$$

$$\omega_{cd} = (10s^{-1})\frac{3in}{5in}\frac{\sin(90° - 30°)}{\sin(180° - 30°)} = 10.4s^{-1}$$

7.3.2.2 Slider-crank mechanism

Example: For the slider-crank mechanism shown in Figure 7-27, bar ab rotates at 1500rpm clockwise. Find the velocity of the piston at point c.

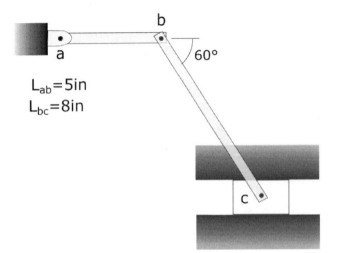

Figure 7-27

Because bar ab rotates about the pin connection at a, the velocity of point b must be vertical as shown in Figure 7-28. Because the piston at c is confined to move in the slot, the velocity of point c must be horizontal. For bar bc we know the direction of the velocity of two points on the bar. Constructing lines perpendicular to v_b and v_c gives the instant center for bar bc at point o. We can calculate the following distances.

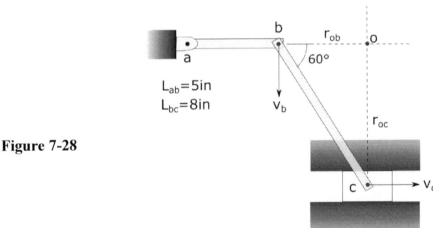

Figure 7-28

$$r_{ob} = 8in\cos 60° = 4in \qquad r_{oc} = 8in\sin 60° = 6.93in$$

We now perform the following series of calculations.

$$v_b = \omega_{ab}r_{ab} = \frac{1500rev}{\min}\left(\frac{2\pi rad}{rev}\right)\left(\frac{1\min}{60s}\right)(5in) = 785in/s$$

$$\omega_{bc} = \frac{v_b}{r_{ob}} = \frac{785in/s}{4in} = 196s^{-1}$$

$$v_c = \omega_{bc} r_{oc} = 196 s^{-1}(6.93 in) = 1361 in / s$$

7.4 Rigid Body Kinetics

7.4.1 Rotation about a fixed axis

Consider a rigid body that rotates about a fixed axis at point q as shown in Figure 7-29.

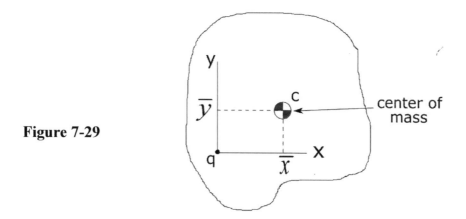

Figure 7-29

center of mass

We establish an x,y-coordinate system with the origin at q so that the center of mass point c has coordinates $\bar{x}$ and $\bar{y}$. In applying Newton's second law for a rigid body, we must deal with the net moment about point q as well as forces. The resulting equations are as follows:

$$\sum F_x = ma_{cx}$$
$$\sum F_y = ma_{cy}$$
$$\sum M_q = [I_c + m(\bar{x}^2 + \bar{y}^2)]\alpha$$

where I_c is the mass moment of inertia about the center of mass (see Table 7-1), a_{cx} and a_{cy} are the components of acceleration of point c, and α is the angular acceleration of the body.

Table 7-1 Mass Moment of Inertia

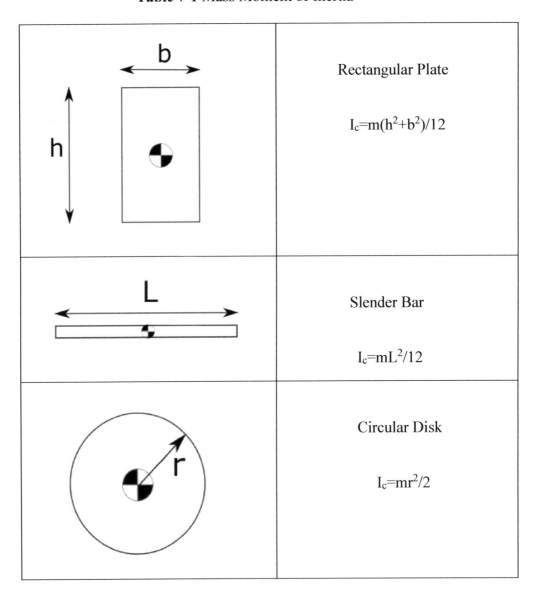

b h	Rectangular Plate $I_c = m(h^2 + b^2)/12$
L	Slender Bar $I_c = mL^2/12$
r	Circular Disk $I_c = mr^2/2$

Example: A 4ft long bar is released from rest in the position shown in Figure 7-30. Find the angular acceleration of the bar if it weighs 10lb.

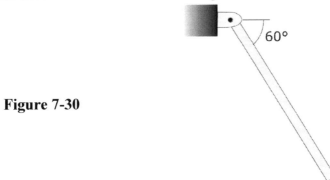

Figure 7-30

We begin by drawing a free body diagram of the bar as shown in Figure 7-31.

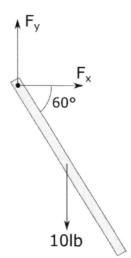

Figure 7-31

For this case we have

$$m = \frac{10}{32.2}\,slug = 0.311\,slug$$

$$I_c = \frac{1}{12}mL^2 = \frac{1}{12}(0.311\,slug)(4\,ft)^2 = 0.414\,slugft^2$$

$$\bar{x} = 2\,ft\cos 60° = 1\,ft$$

$$\bar{y} = -(2\,ft)\sin 60° = -1.73\,ft$$

$$\sum M_q = [I_c + m(\bar{x}^2 + \bar{y}^2)]\alpha \rightarrow -(10lb)(1\,ft) = \{0.414\,slugft^2 + 0.311\,slug[(1\,ft)^2 + (-1.73\,ft)^2]\}\alpha$$

Solving gives

$$\alpha = -6.04s^{-2}$$

7.4.2 Work-energy analysis

Recall the following work-energy relation for a particle:

$$d\sum_{i=1}^{n}(F_i\cos\theta_i) = \frac{1}{2}m(v_f^2 - v_s^2) + \frac{1}{2}k(\delta_f^2 - \delta_s^2) + mg(h_f - h_s)$$

We have the following relation for rigid bodies that includes the kinetic energy due to rotation ($I_c\omega^2/2$).

$$d\sum_{i=1}^{n}(F_i\cos\theta_i)=\frac{1}{2}m(v_{cf}^2-v_{cs}^2)+\frac{1}{2}k(\delta_f^2-\delta_s^2)+mg(h_{cf}-h_{cs})+\frac{1}{2}I_c(\omega_f^2-\omega_s^2)$$

where the subscript c refers to the center of mass point, s refers to the starting position, and f refers to the finishing position..

Example: A bar is released from rest in the horizontal position as shown in Figure 7-32. If the bar weighs 10lb, find the angular velocity when it reaches the vertical position.

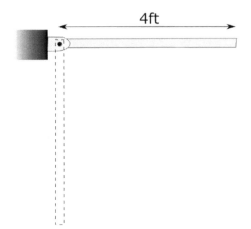

Figure 7-32

From the previous example, we have

$$m=0.311slug$$
$$I_c=0.414slugft^2$$

Since the bar rotates about the hinge, we have

$$v_c=\omega r=\omega(2\,ft)$$

The only force that does work is gravity which is accounted for by the potential energy term mgh_c. We have

$$h_{cs}=0 \qquad h_{cf}=-1\,ft$$

The work-energy equation for this case is

$$0=\frac{1}{2}mv_{cf}^2+mgh_{cf}+\frac{1}{2}I_c\omega^2\rightarrow 0=\frac{1}{2}(0.311slug)[(2\,ft)\omega]^2+10lb(-2\,ft)+\frac{1}{2}(0.414slugft^2)\omega^2$$

Solving gives

$$\omega = 24.1 s^{-1}$$

7.5 Vibrations

7.5.1 Free vibration with damping

Vibration problems are typically idealized as an oscillating mass m connected to a spring with stiffness k and a damper with damping c as shown in Figure 7-33.

Figure 7-33

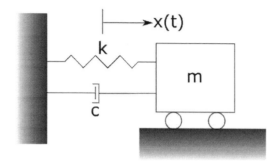

The application of Newton's second law gives the following equation of motion:

$$m\ddot{x} + c\dot{x} + kx = 0$$

where x is the displacement of the mass. This equation can also be written as

$$\ddot{x} + 2\varsigma\omega_n\dot{x} + \omega_n^2 x = 0$$

where $\omega_n = (k/m)^{1/2}$ is the undamped natural frequency, and $\varsigma = c/(2m\omega_n)$ is the damping ratio. If the mass has an initial displacement x_o and an initial velocity v_o, the solution is

$$x(t) = Ae^{-\varsigma\omega_n t} \sin(\omega_d t + \phi)$$

where

$$A = \sqrt{x_o^2 + (v_o + x_o\varsigma\omega_n)^2 / \omega_d^2}$$

$$\phi = \tan^{-1}\left(\frac{x_o}{(x_o\varsigma\omega_n + v_o)/\omega_d}\right)$$

and $\omega_d = \omega_n(1-\varsigma^2)^{1/2}$ is the damped natural frequency.

A plot of displacement versus time is shown in Figure 7-34.

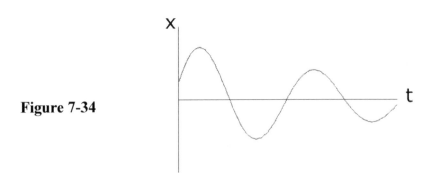

Figure 7-34

The period $T=2\pi/\omega_d$ is the time between peaks.

Example: The mass in the system shown in Figure 7-35 starts from rest with an initial displacement of 3cm. Find how much the displacement has decreased after three cycles of oscillation.

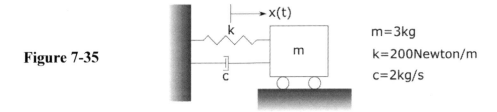

Figure 7-35

m=3kg
k=200Newton/m
c=2kg/s

The starting displacement (at t=0) can be written as

$$x_s = A\sin(\phi)$$

The final displacement is the value of x when $t=3T=6\pi/\omega_d$; i.e.,

$$x_f = Ae^{-\varsigma\omega_n(6\pi)/\omega_d}\sin(6\pi + \phi)$$

Consider the ratio of the final displacement to initial displacement

$$\frac{x_f}{x_s} = \frac{Ae^{-\varsigma\omega_n(6\pi)/\omega_d}\sin(6\pi + \phi)}{A\sin(\phi)} = e^{-6\pi\varsigma/\sqrt{1-\varsigma^2}}$$

where we have used the relation $\sin(2n\pi + \phi) = \sin\phi$

For this case

$$\omega_n = \sqrt{\frac{k}{m}} = \sqrt{\frac{200Newton/m}{3kg}\left(\frac{1kgm/s^2}{1Newton}\right)} = 8.16s^{-1}$$

$$\varsigma = \frac{c}{2m\omega_n} = \frac{2kg/s}{2(3kg)(8.16s^{-1})} = 0.04$$

$$\frac{x_f}{x_s} = e^{-6\pi\varsigma/\sqrt{1-\varsigma^2}} = e^{-6\pi(0.04)/\sqrt{1-(0.04)^2}} = 0.47$$

Therefore, the displacement has decreased by 53%.

7.5.2 Free vibration without damping

Consider the previous case without damping but with the mass suspended vertically and with gravity accounted for as shown in Figure 7-36.

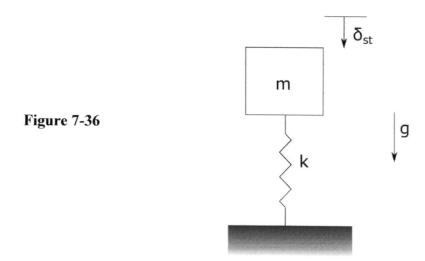

Figure 7-36

The static deflection δ_{st}=mg/k is that due to gravity acting alone. The additional oscillatory deflection is x and is given by

$$x(t) = x_o \cos \omega_n t + \frac{v_o}{\omega_n} \sin \omega_n t$$

As before,

$$\omega_n = \sqrt{\frac{k}{m}} = \sqrt{\frac{g}{\delta_{st}}}$$

Example: A 100lb weight is attached to the end of the cantilever beam shown in Figure 7-37. Find the natural frequency of the system ignoring the mass of the beam.

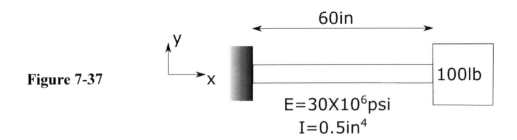

Figure 7-37

The beam behaves like a spring in the sense that it experiences a deflection that is proportional to the load applied at the end. Let's find δ_{st}. From the beam deflection table, we have

$$\delta_{st} = \frac{WL^3}{3EI} = \frac{100lb(60in)^3}{3(30 \times 10^6\ psi)(0.5in^4)} = 0.48in$$

Then

$$\omega_n = \sqrt{\frac{32.2\ ft/s^2}{0.48in}\left(\frac{12in}{1ft}\right)} = 28.4s^{-1}$$

8. SYSTEMS AND CONTROLS

8.1 Laplace Transform

The Laplace transform allows us to convert differential equations with time as the independent variable into algebraic equations. The Laplace transform of the function f(t) is defined as

$$F(s) = L\{f(t)\} = \int_0^\infty f(t)e^{-st}dt$$

The Laplace transforms of some elementary functions are given in Table 8-1.

Table 8-1

f(t)	F(s)
$e^{-\alpha t}$	$\dfrac{1}{s+\alpha}$
$e^{-\alpha t}\sin\beta t$	$\dfrac{\beta}{(s+\alpha)^2+\beta^2}$
$e^{-\alpha t}\cos\beta t$	$\dfrac{s+\alpha}{(s+\alpha)^2+\beta^2}$

The Laplace transform of the first derivative of f(t) is

$$L\left\{\frac{df}{dt}\right\} = sF(S) - f(0)$$

The Laplace transform of the second derivative of f(t) is

$$L\left\{\frac{d^2 f}{dt^2}\right\} = s^2 F(S) - sf(0) - \dot{f}(0)$$

Example: Find the solution to the following differential equation:

$$\frac{dx}{dt} = -2x \qquad x(0) = 10$$

Taking the Laplace transform of each term in the equation gives

$$sX(s) - 10 = -2X(s)$$

Solving for X(s) gives

$$X(s) = \frac{10}{s+2}$$

Using Table 8-1 to find the inverse transform gives x(t) as

$$x(t) = 10e^{-2t}$$

It is interesting to compare this solution technique to the one from Section 1.2.

8.2 Transfer Function

A dynamic system can be thought of as having an input and an output. The transfer function is defined as the ratio of the Laplace transform of the output to the Laplace transform of the input; i.e.,

$$\text{Transfer Function} \qquad T(s) = \frac{L\{\text{Output}\}}{L\{\text{Input}\}}$$

A classic example of a second order dynamic system is the spring-mass-damper system shown in Figure 8-1.

Figure 8-1

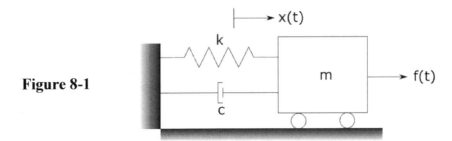

The differential equation for the system (see Section 7.5.1) is

$$m\frac{d^2x}{dt^2} + c\frac{dx}{dt} + kx = f(t) \qquad x(0) = \dot{x}(0) = 0$$

In this case f(t) is the input, and x(t) is the output. Taking the Laplace transform of each term gives

$$ms^2X(s) + csX(s) + kX(s) = F(s)$$

Solving for X(s) gives

$$X(s) = \frac{F(s)}{ms^2 + cs + k} = \frac{\frac{1}{m}F(s)}{s^2 + 2\omega_n\varsigma s + \omega_n^2}$$

where $\omega_n = \sqrt{k/m}$ is the undamped natural frequency, and $\varsigma = c/(2m\omega_n)$ is the damping ratio.

In this case, the transfer function is

$$T(s) = \frac{1}{ms^2 + cs + k}$$

The operation of a transfer function can be represented pictorially in a block diagram with input F(s) entering and output X(s) exiting as shown in Figure 8-2 for the previous case.

Figure 8-2   F(s) ⟶ | T(s) | ⟶ X(s)

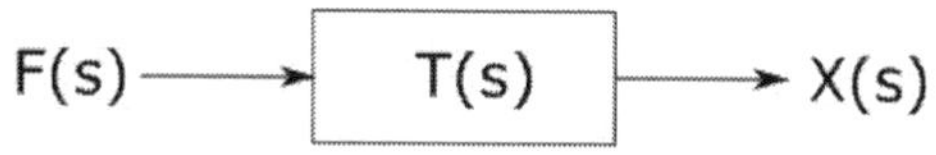

The characteristic equation for the system can be obtained by taking the denominator in the transfer function and setting it equal to zero. The characteristic equation for the second order system is

$$s^2 + 2\omega_n\varsigma s + \omega_n^2 = 0$$

Comparing this to the characteristic equation in Section 1.2, we observe that they are effectively identical. The roots of this equation are complex conjugates.

$$s = -\omega_n\varsigma \pm j\omega_n\sqrt{1-\varsigma^2} \qquad \text{where} \quad j = \sqrt{-1}$$

8.3 Control Systems

A closed-loop feedback system is often used to control some variable c(t) given a regulating input r(t). The block diagram for a simple case is shown in Figure 8-3.

Figure 8-3

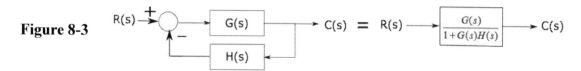

The overall transfer function for the system is

$$T(s) = \frac{G(s)}{1+G(s)H(s)}$$

G(s) is the open-loop transfer function, and H(s) is the feedback transfer function.

A system may have multiple transfer functions. Those in parallel in a block diagram add as shown in Figure 8-4.

Figure 8-4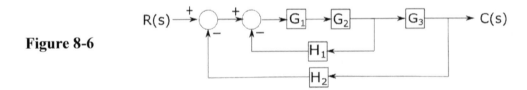

Those in series multiply as shown in Figure 8-5.

Figure 8-5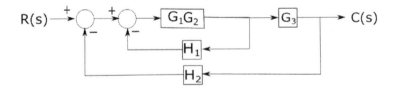

The overall transfer function for complex block diagrams can be determined by applying these principles one at a time.

Example: Find the overall transfer function for the block diagram in Figure 8-6.

Figure 8-6

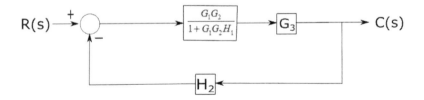

First combine G₁ and G₂.

Now reduce the loop with H₁.

Now account for G_3.

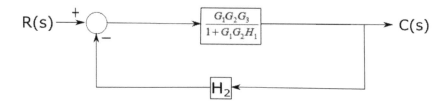

Now reduce the outer loop with H_2.

$$R(s) \longrightarrow \boxed{T(s)} \longrightarrow C(s)$$

$$T(s) = \dfrac{\dfrac{G_1 G_2 G_3}{1 + G_1 G_2 H_1}}{1 + \dfrac{G_1 G_2 G_3}{1 + G_1 G_2 H_1} H_2} = \dfrac{G_1 G_2 G_3}{1 + G_1 G_2 H_1 + G_1 G_2 G_3 H_2}$$

8.4 Stability

In a feedback system, there is the possibility of the system becoming unstable where the response increases over time without limit. The system is stable if the real parts of the roots of the characteristic equation are negative. The Routh-Hurwitz criterion can be used to determine this. A tabular version can be used for an nth-degree polynomial of the form

$$a_n s^n + a_{n-1} s^{n-1} + \ldots + a_1 s + a_0 = 0$$

We form the table in Table 8-2.

Table 8-2

a_n	a_{n-2}	a_{n-4}	....
a_{n-1}	a_{n-3}	a_{n-5}	....
b_1	b_2	b_3	....
c_1	c_2	....	....
d_1	....	....	....
....	....	....	....

where

$$b_1 = \frac{a_{n-1}a_{n-2} - a_n a_{n-3}}{a_{n-1}} \qquad b_2 = \frac{a_{n-1}a_{n-4} - a_n a_{n-5}}{a_{n-1}} \qquad b_3 = \frac{a_{n-1}a_{n-6} - a_n a_{n-7}}{a_{n-1}}$$

$$c_1 = \frac{b_1 a_{n-3} - a_{n-1}b_2}{b_1} \qquad c_2 = \frac{b_1 a_{n-5} - a_{n-1}b_3}{b_1} \qquad d_1 = \frac{c_1 b_2 - b_1 c_2}{c_1}$$

All the terms in the first column of the table must be of the same sign and non-zero for all the roots to have negative real parts.

Example: Determine if the system with the following characteristic equation is stable.

$$s^4 - 4s^3 + 3s^2 + 14s + 26 = 0$$

We set up the table as shown below

$$b_1 = \frac{a_{n-1}a_{n-2} - a_n a_{n-3}}{a_{n-1}} = \frac{-4(3) - 1(14)}{-4} = 13/2$$

$$b_2 = \frac{a_{n-1}a_{n-4} - a_n a_{n-5}}{a_{n-1}} = \frac{-4(26) - 0}{-4} = 26$$

$$b_3 = \frac{a_{n-1}a_{n-6} - a_n a_{n-7}}{a_{n-1}} = \frac{-4(0) - 1(0)}{-4} = 0$$

$$c_1 = \frac{b_1 a_{n-3} - a_{n-1}b_2}{b_1} = \frac{13/2(14) - (-4)(26)}{13/2} = 30$$

$$c_2 = \frac{b_1 a_{n-5} - a_{n-1}b_3}{b_1} = \frac{13/2(0) - (-4)(0)}{13/2} = 0$$

$$d_1 = \frac{c_1 b_2 - b_1 c_2}{c_1} = \frac{30(26) - 13/2(0)}{30} = 26$$

1	3	26	
-4	14		
6.5	26		
30			

Not all the terms have the same sign in the first column. Therefore, the system is unstable.

9. MATERIALS

9.1 Mechanical Properties of Materials

9.1.1 Stress-strain behavior

Some of the most important mechanical properties of a material can be determined from a stress-strain diagram which is obtained from a uniaxial tension test. An idealized version of one of these diagrams typical of metals is shown in Figure 9-1.

Figure 9-1

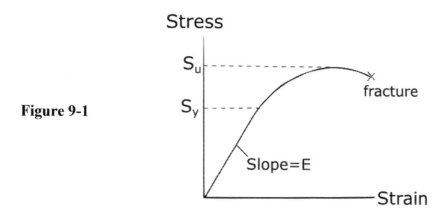

The stress is in terms of force per unit area, and the strain is in terms of elongation per unit length. The initial part of the curve is linear with slope E (Young's modulus or elastic modulus) and represents purely elastic response. When the stress exceeds the yield point S_y, the curve becomes nonlinear with a considerably smaller slope. The maximum stress that can be sustained by the material is the ultimate stress S_u.

Some materials do not exhibit a distinct yield point (see Figure 9-2). To estimate the yield stress for this case, a line starting at a strain of 0.002 running parallel to the initial slope of the stress-strain curve (slope=E) is constructed (dashed line). The yield stress is designated as the point where the dashed line intersects the stress-strain curve. This technique is called the 0.2 percent offset method.

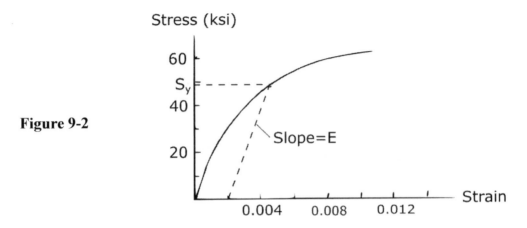

Figure 9-2

9.1.2 Elastic plastic behavior

If the material is loaded beyond the yield point and then the load is removed, the unloading stress-strain curve does not retrace the original loading curve, but instead follows a straight line parallel to the original elastic part of the stress-strain curve as shown in Figure 9-3.

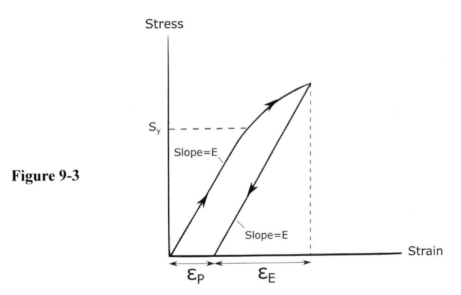

Figure 9-3

When the stress reaches zero, there is a permanent plastic strain ε_P in the material. The total strain at the maximum load point consists of an elastic part ε_E and a plastic part ε_P;

$$\varepsilon_{total} = \varepsilon_E + \varepsilon_P \quad \text{with} \quad \varepsilon_E = \sigma / E$$

Example: A material with the stress-strain curve shown in Figure 9-4 is subjected to a stress of 55ksi. Find the permanent plastic strain that remains after the load is removed.

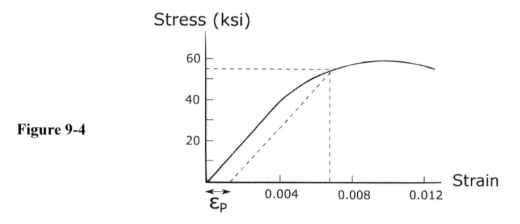

Figure 9-4

From the Figure we observe that when the stress is 55ksi, the strain is 0.0067. We calculate the elastic modulus as

$$E = S_y / \varepsilon_y = 40{,}000\, psi / 0.004 = 10 \times 10^6\, psi$$

The permanent plastic strain is

$$\varepsilon_P = \varepsilon_{total} - \sigma / E = 0.0067 - 55{,}000\, psi / 10 \times 10^6\, psi = 0.0012$$

9.1.3 Creep behavior

Materials at high temperature will often deform over time even though the stress remains constant. A typical plot of strain versus time at constant stress for this situation is shown in Figure 9-5.

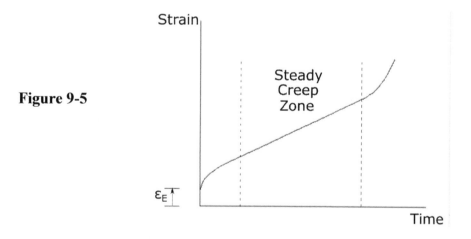

Figure 9-5

The behavior of that part of the curve to the left of the two dashed lines is called primary creep. The behavior for that part of the curve between the two dashed lines is called secondary creep (or steady creep). The curve in the steady creep region is effectively a straight line. In this region, the strain rate is independent of time and can be expressed as

$$\frac{d\varepsilon_c}{dt} = A\sigma^n e^{Q/(RT)}$$

where

A=pre-exponent constant
n=stress sensitivity
Q=activation energy

are material properties, and

σ=stress
T=absolute temperature
R=ideal gas law constant

The initial strain at t=0 is simply the elastic strain given by

$$\varepsilon_E = \sigma / E$$

In an analysis, the primary creep phase is often ignored, and the steady creep phase is assumed to hold for all time t>0. Thus, at any time, the total strain is

$$\varepsilon_{total} = \sigma / E + At\sigma^n e^{Q/(RT)}$$

Example: A bar with a diameter of 20cm and a length of 0.5m is subjected to a tensile load of 25,000Newton and a temperature of 600°C. Find the elongation of the bar after 4hr if the bar has the following properties: E=50GPa, A=4.5x10^{-34}m^{6}Newton^{-3}hr^{-1}, n=3, Q=130,000J/mol.

We calculate the stress as

$$\sigma = \frac{F}{Area} = \frac{25,000 Newton}{\pi(0.01m)^2} = 80 \times 10^6 \, Newton / m^2$$

The strain is

$$\varepsilon_{total} = \frac{80 \times 10^6 \, Newton / m^2}{50 \times 10^9 \, Newton / m^2}$$
$$+ (4.5 \times 10^{-34} m^6 Newton^{-3} hr^{-1})(4hr)(80 \times 10^6 \, Newton / m^2)^3 e^{130,000J/mol/[8.314J/mol°K(873°K)]}$$
$$\varepsilon_{total} = 0.0569$$

The elongation is

$$\delta = \varepsilon L = 0.0569(0.5m) = 0.0285m$$

9.2 Hardness

A hardness test provides a way to obtain an estimate of a material's ultimate strength without a stress-strain curve. This typically involves pressing an indenter into the surface of the material and measuring the depth of the penetration. From this, a hardness number is determined which can be correlated with ultimate strength.

The ultimate strength of steel in psi can be estimated from

$$S_u \approx 500 H_B$$

where H_B is the Brinnell hardness.

9.3 Phase Diagrams

Metallic structural materials are typically not pure metals but are alloys containing impurity atoms of another element to produce an improvement in properties (for example, carbon in iron to form steel). For steel and cast iron, we can represent the various combinations of iron with carbon in a phase diagram as shown in Figure 9-6 where the horizontal axis gives the amount of carbon, and the vertical axis gives the temperature. For temperatures below 1420°F, we have a combination of ferrite (α-iron) and cementite (Fe_3C, iron carbide). Carbon concentrations above 6.7% are pure cementite and are of no interest. The primary phases shown are ferrite (α-iron, with a body-centered cubic crystal structure), austenite (γ-iron with a face-centered cubic structure), δ-ferrite (δ-iron with body-centered cubic structure), cementite, and liquid (L). Iron-carbon combinations with carbon concentrations below 2% are considered to be steel while iron-carbon combinations with carbon concentrations between 2% and 6.7% are considered to be cast iron.

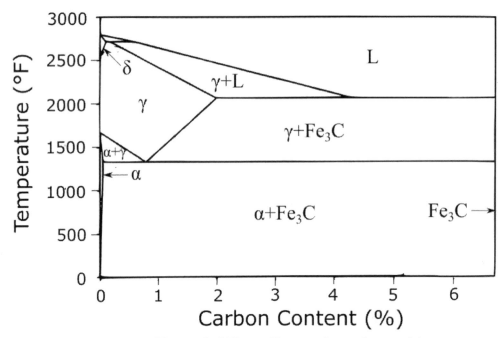

Figure 9-6 Phase diagram for carbon and iron

Given the carbon concentration and temperature, the amount of each phase that is present can be determined using the lever rule. This rule is illustrated in Figure 9-7 for a material with phases alpha (α), beta (β), and liquid (L) that has a concentration X_0 and temperature T_0 on the phase diagram.

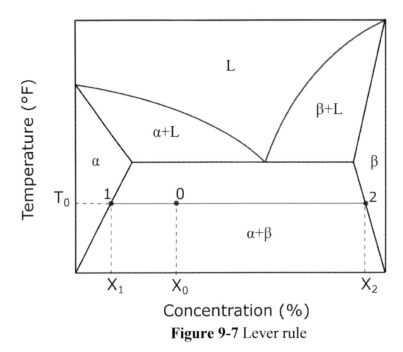

Figure 9-7 Lever rule

We want to find the amount of phases alpha and beta at point 0. First, a tie -line connecting points 1, 0 and 2 is constructed. Next, we seek the fraction of alpha F_α at point 1 that would balance the fraction of beta F_β at point 2 if point 0 were the pivot point for a lever running between points 1 and 2. A moment balance requires the following.

$$F_\alpha L_{10} = F_\beta L_{02} \qquad \text{or} \qquad \frac{F_\alpha}{F_\beta} = \frac{L_{02}}{L_{10}} = \frac{X_2 - X_0}{X_0 - X_1}$$

This relationship together with

$$F_\alpha + F_\beta = 100\%$$

allows us to calculate F_α and F_β.

<u>Example</u>: Determine the amount of austenite and cementite in cast iron with a carbon content of 3% at 1500°F.

The situation is designated as point 0 on the iron-carbon phase diagram as shown in Figure 9-8.

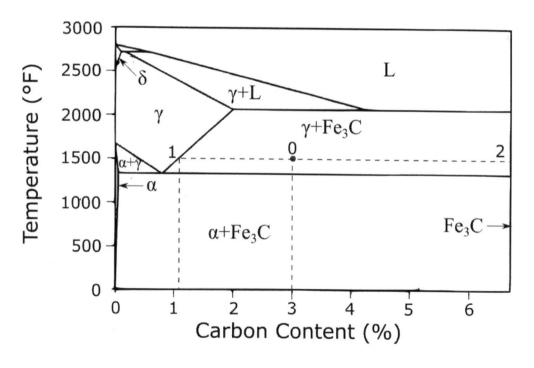

Figure 9-8

From the tie line, we get $C_1 = 1.1\%$ and $C_2 = 6.7\%$. Therefore,

$$\frac{F_\gamma}{F_{Fe3C}} = \frac{X_2 - X_0}{X_0 - X_1} = \frac{6.7\% - 3\%}{3\% - 1.1\%} = 2.47 \qquad \text{and} \qquad F_\gamma + F_{Fe3C} = 100\%$$

This gives F_γ=71.2% and F_{Fe3C}=28.8%.

9.4 Heat Treatment of Steel

The properties of steel are affected by heat treatment. This involves heating the steel to a high enough temperature to produce austenite followed by rapid cooling. This has the effect of increasing the yield strength and reducing ductility. Cold working the steel has a similar effect. Ductility can be restored at the expense of reduced yield strength through a tempering process that involves reheating the steel and cooling it more slowly. This process also relieves residual stresses that may be present.

9.5 Composite Materials

Fiber reinforced plastics are often used in high-performance structural applications where high stiffness and high strength coupled with low weight are required. These composite materials consist of high-strength fibers embedded in low strength matrix material as shown in Figure 9-9.

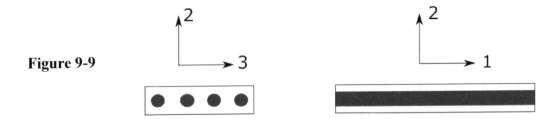

Figure 9-9

The properties of this composite material are a mixture of the properties of the fibers and the matrix. The elastic modulus parallel to the fibers (1-direction) is

$$E_{c1} = E_f V_f + E_m V_m$$

where
E_f = elastic modulus of fiber
E_m = elastic modulus of matrix
V_f = volume fraction of fiber
V_m = volume fraction of fiber

with $V_f + V_m = 1$

The elastic modulus perpendicular to the fibers is

$$E_{c2} = \frac{E_f E_m}{E_f V_m + E_m V_f}$$

For brittle fibers the tensile strength of the composite parallel to the fibers (1-direction) is

$$\sigma_{c1} = \sigma_f \left(V_f + \frac{E_m}{E_f} V_m \right)$$

where σ_f is the fiber strength.

For ductile fibers the tensile strength of the composite parallel to the fibers (1-direction) is

$$\sigma_{c1} = \sigma_f V_f + \sigma_m V_m$$

where σ_m is the matrix strength.

Example: A carbon/epoxy composite has a fiber volume fraction of 70%. With $E_f=27 \times 10^6$psi and $E_m=0.5 \times 10^6$psi, find the elastic moduli of the composite.

For this case we have $V_f=0.7$ and $V_m=0.3$. Then,

$$E_{c1} = E_f V_f + E_m V_m = 27 \times 10^6 \, psi(0.7) + 0.5 \times 10^6 \, psi(0.3) = 19.1 \times 10^6 \, psi$$

$$E_{c2} = \frac{E_f E_m}{E_f V_m + E_m V_f} = \frac{(27 \times 10^6 \, psi)(0.5 \times 10^6 \, psi)}{(27 \times 10^6 \, psi)(0.3) + (0.5 \times 10^6 \, psi)(0.7)} = 1.6 \times 10^6 \, psi$$

We note the low stiffness perpendicular to the fibers. To eliminate this problem, structures made of composite materials are typically composed of many layers with alternating directions of fibers.

10. MACHINE DESIGN

10.1 Failure of Ductile Materials

10.1.1 Maximum distortion energy failure criterion

In general, parts are designed so that they do not experience yielding. The yield stress is determined from a uniaxial tension test where there is only one normal stress. Determining when a material will yield under a complex state of stress requires a general yield criterion. One that is valid for most metals is the maximum distortion energy criterion (Von Mises) which states that yielding will occur when the following condition is met

$$\frac{1}{\sqrt{2}}[(\sigma_x - \sigma_y)^2 + (\sigma_x - \sigma_z)^2 + (\sigma_y - \sigma_z)^2 + 6(\tau_{xy}^2 + \tau_{xz}^2 + \tau_{yz}^2)]^{1/2} = S_y$$

The expression on the left side of the equation above is called the Von Mises stress. For the case of plane stress (σ_z=0, τ_{xz}=0, τ_{yz}=0), this criterion becomes

$$[\sigma_x^2 - \sigma_x\sigma_y + \sigma_y^2 + 3\tau_{xy}^2]^{1/2} = S_y$$

10.1.1 Maximum shear stress failure criterion

A slightly more conservative criterion that may be easier to apply in certain circumstances is the maximum shear stress criterion (Tresca). This criterion states that yield will occur when

$$\tau_{max} = S_y/2$$

The maximum shear stress at a point is equal to one half the difference between the maximum and minimum normal stresses. Thus, this criterion becomes

$$\sigma_{max} - \sigma_{min} = S_y$$

Example: Determine the factor of safety for a material with S_y=40ksi that is subject to the plane stress state σ_x=20ksi, σ_y=10ksi, τ_{xy}=5ksi .

Let's calculate the Von Mises stress first

$$[\sigma_x^2 - \sigma_x\sigma_y + \sigma_y^2 + 3\tau_{xy}^2]^{1/2} = [(20ksi)^2 - (20ksi)(10ksi) + (10ksi)^2 + 3(5ksi)^2]^{1/2} = 19.36ksi$$

The maximum distortion energy criterion gives the safety factor as

$$S.F. = \frac{40ksi}{19.36ksi} = 2.06$$

Now let's consider the maximum shear stress criterion. First, we calculate the principal stresses in the x,y-plane (See Section 6.2).

$$\sigma_{1,2} = \frac{\sigma_x + \sigma_y}{2} \pm \sqrt{\left(\frac{\sigma_x - \sigma_y}{2}\right)^2 + \tau_{xy}^2} = \frac{20ksi + 10ksi}{2} \pm \sqrt{\left(\frac{20ksi - 10ksi}{2}\right)^2 + (5ksi)^2} = 22.07ksi, 7.93ksi$$

The maximum normal stress is 22.07ksi. The minimum normal stress is **not** 7.93ksi. We need to consider the z-direction as well as x and y. The minimum normal stress is $\sigma_z = 0$. Therefore,

$$\sigma_{max} - \sigma_{min} = 22.07ksi - 0 = 22.07ksi$$

The safety factor is

$$S.F. = \frac{40ksi}{22.07ksi} = 1.81$$

10.2 Failure of Brittle Materials

10.2.1 Maximum normal stress criterion

Brittle materials typically have no discernible yield point. For this case we can use the maximum normal stress criterion. To apply this criterion, we calculate the principal stresses (see section 6.2). Failure occurs when the largest of these equals the ultimate strength of the material S_U.

$$\sigma_{max} = S_U$$

Using the equation from section 6.2 for plane stress, this can be rewritten as

$$\frac{\sigma_x + \sigma_y}{2} + \sqrt{(\frac{\sigma_x - \sigma_y}{2})^2 + \tau_{xy}^2} = S_U$$

10.2.2 Fracture mechanics

If information about the presence of cracks is available, a fracture mechanics approach may be used. An elastic analysis of the stress field around a crack indicates a singularity in stress at the crack tip. A crack propagation criterion can be developed based on the magnitude of this singularity which can be expressed in terms of the stress intensity factor K_I. Fracture is predicted to occur when the stress intensity factor reaches the critical value K_{IC} (also called the fracture toughness) which is a material property. Fracture occurs when,

$$K_I = K_{IC}$$

Formulas for K_I have been determined for a number of cases. Consider a large plate under tension with a central crack with length 2a as shown in Figure 10-1.

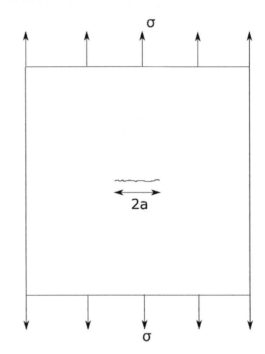

Figure 10-1

The stress intensity factor for this case is

$$K_I = \sigma\sqrt{\pi a}$$

Consider a large plate under tension with an edge crack of length a as shown in Figure 10-2.

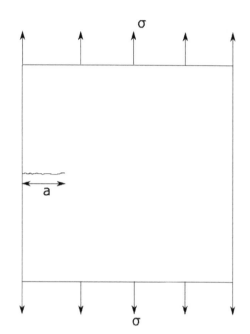

Figure 10-2

The stress intensity factor for this case is

$$K_I = 1.1\sigma\sqrt{\pi a}$$

Example: A 10in x 10in plate that is 0.25in thick carries a tensile load of 10,000lb. If the fracture toughness of the material is K_{IC}=10,000psiin$^{1/2}$, find the maximum allowable length of an edge crack.

First, we calculate the stress in the plate ignoring the presence of any cracks.

$$\sigma = \frac{F}{A} = \frac{10,000lb}{10in(0.25in)} = 4,000\,psi$$

Setting

$$1.1\sigma\sqrt{\pi a} = K_{IC}$$

and solving for a gives

$$a = \frac{K_{IC}^2}{\pi(1.1\sigma)^2} = \frac{(10,000\,psiin^{1/2})^2}{\pi[1.1(4,000\,psi)]^2} = 1.64in$$

10.3 Fatigue Failure

10.3.1 Cyclic loading

Rotating machinery often contain parts subjected to cyclic load as shown in Figure 10-3.

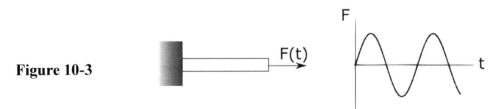

Figure 10-3

Material subjected to this type of loading will often fail at a stress that is significantly lower than that needed to cause failure under constant load. Failure does not usually occur until the part experiences a certain number of load cycles. The number of cycles N to cause failure decreases as the stress S_f increases. For many materials a plot of failure stress versus number of load cycles looks like that shown in Figure 10-4.

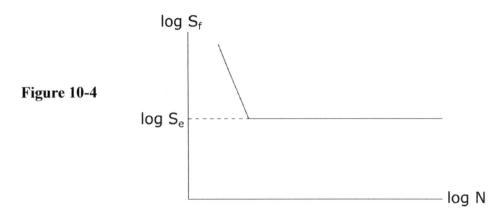

Figure 10-4

For stress levels below S_e, the part will not fail regardless of the number of cycles. This value of stress for which the part will have infinite life is called the endurance limit.

10.3.2 Modifying factors

The endurance limit S_e on the fatigue curve is for the case of ideal conditions. The effective endurance limit S_{eff}, taking into account realistic conditions, is typically smaller and is determined by lowering S_e by a modifying factor. The effective endurance limit is

$$S_{eff} = K_{surf}K_{size}K_{temp}K_{cor}S_e$$

where K_{surf}= surface effects factor (roughness lowers strength)
K_{size}=size factor (increased size lowers strength)
K_{temp}=temperature factor (high temperature lowers strength)
K_{cor}= corrosion factor (corrosion lowers strength)

10.3.3 Combined constant and cyclic loading

Consider the case of a cyclic stress σ_a superimposed on a constant stress σ_m (mean stress) so that the stress versus time curve is as shown in Figure 10-5.

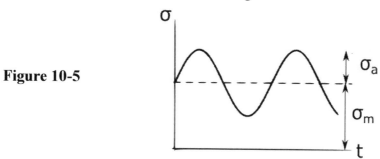

Figure 10-5

If $\sigma_m=0$ (cyclic loading only), we would expect the stress for infinite life to be S_e. If $\sigma_a=0$ (constant load only), we would expect the stress for infinite life to be the ultimate strength S_U. For stress situations between these two extremes, we draw a straight line (Goodman line) between the two extreme points of the plot of σ_a for failure versus σ_m for failure as shown in Figure 10-6

.

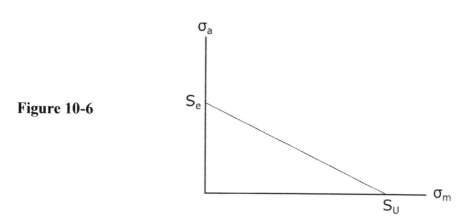

Figure 10-6

Stress states below the line are safe. We can express the failure criterion as

$$\frac{\sigma_a}{S_e} + \frac{\sigma_m}{S_U} = 1$$

There is a slightly different version of this criterion (Soderberg) where S_U is replaced by the yield strength S_y. Failure occurs when the following conditions are satisfied.

$$\frac{\sigma_a}{S_e} + \frac{\sigma_m}{S_Y} = 1$$

Example: A 1in diameter bar carries a weight of 5,000lb as shown in Figure 10-7. If S_e=20,000psi and S_y=32,000psi, use the Soderberg criterion to find the allowable cyclic load for infinite life of the bar.

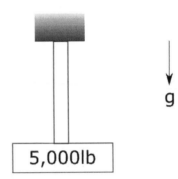

Figure 10-7

g

5,000lb

First, we calculate σ_m as

$$\sigma_m = \frac{W}{A} = \frac{5,000lb}{\pi(1in)^2/4} = 6,366\,psi$$

solving σ_a for from the Soderberg criterion gives

$$\sigma_a = S_e\left(1-\frac{\sigma_m}{S_y}\right) = 20,000\,psi\left(1-\frac{6,366\,psi}{32,000\,psi}\right) = 16,021\,psi$$

The force associated with this stress is

$$F_a = \sigma_a A = 16,021\,psi[\pi(1in)^2/4] = 12,583lb$$

10.4 Pressurized Cylinders

10.4.1 Thick cylinder with internal pressure

Consider a thick cylinder with inside pressure p_i acting at r_i as shown in Figure 10-8.

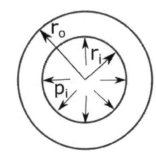

Figure 10-8

There will be a normal stress in the radial direction σ_r and a normal stress in the circumferential direction σ_θ as follows:

$$\sigma_r = \frac{p_i r_i^2}{r_o^2 - r_i^2}(1 - \frac{r_o^2}{r^2})$$

$$\sigma_\theta = \frac{p_i r_i^2}{r_o^2 - r_i^2}(1 + \frac{r_o^2}{r^2})$$

Note that the stresses depend on radial position r but not on angular position θ. The largest stresses occur at the inner radius r_i and are given by

$$\sigma_r = -p_i \qquad \sigma_\theta = \frac{p_i(r_o^2 + r_i^2)}{r_o^2 - r_i^2} = \sigma_{max}$$

The radial displacement u is given by

$$u = r\varepsilon_\theta = \frac{r}{E}(\sigma_\theta - v\sigma_r)$$

If the cylinder has end caps, there will be a normal stress in the axial direction given by

$$\sigma_a = \frac{p_i r_i^2}{r_o^2 - r_i^2}$$

Example: A pressure vessel with r_i=5in and r_o=6in and with end caps is subjected to an internal pressure of p_i=3,000psi . Find the safety factor using the maximum distortion energy criterion if S_y=40,0000psi.

First, we calculate the stresses at the inner radius.

$$\sigma_r = -p_i = -3,000psi$$

$$\sigma_\theta = \frac{p_i(r_o^2 + r_i^2)}{r_o^2 - r_i^2} = \frac{3,000psi[(6in)^2 + (5in)^2]}{(6in)^2 - (5in)^2} = 16,637psi$$

$$\sigma_a = \frac{p_i r_i^2}{r_o^2 - r_i^2} = \frac{3,000psi(5in)^2}{(6in)^2 - (5in)^2} = 6,818psi$$

We now calculate the Von Mises stress using cylindrical coordinates with zero shear stresses.

$$\frac{1}{\sqrt{2}}[(\sigma_r - \sigma_\theta)^2 + (\sigma_r - \sigma_a)^2 + (\sigma_\theta - \sigma_a)^2]^{1/2}$$

$$= \frac{1}{\sqrt{2}}[(-3,000\,psi - 16,637\,psi)^2 + (-3,000\,psi - 6,818\,psi)^2 + (16,637\,psi - 6,818\,pis)^2]^{1/2}$$

$$= 17,006\,psi$$

The safety factor is

$$S.F. = \frac{40,000\,psi}{17,006\,psi} = 2.35$$

10.4.2 Thick cylinder with external pressure

Consider the previous case with outer pressure p_o at r_o as shown in Figure 10-9.

Figure 10-9

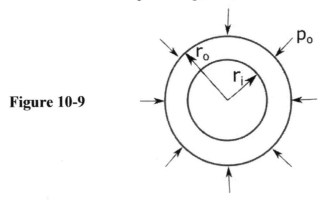

The stresses in this case are

$$\sigma_r = \frac{-p_o r_0^2}{r_o^2 - r_i^2}(1 - \frac{r_i^2}{r^2})$$

$$\sigma_\theta = \frac{-p_o r_0^2}{r_o^2 - r_i^2}(1 + \frac{r_i^2}{r^2})$$

The largest stresses occur at the inner radius r_i and are given by

$$\sigma_r = 0 \qquad \sigma_\theta = \frac{-2p_o r_o^2}{r_o^2 - r_i^2} = \sigma_{max}$$

10.4.3 Thin cylinders

If the cylinder thickness is than 10% of the inner radius for an internally pressurized cylinder, we can use the thin-walled approximation

$$\sigma_r = -p_i \qquad \sigma_\theta = p_i \frac{r_m}{t}$$

where $r_m = (r_o + r_i)/2$. If the cylinder is capped at the ends, there will also be an axial stress

$$\sigma_a = \frac{1}{2} p \frac{r_m}{t}$$

10.5 Interference Fits

Consider two concentric cylinders as shown in Figure 10-10 with the inner cylinder (with $E = E_i$ and $v = v_i$) having inside radius a and outside radius b and the outer cylinder (with $E = E_o$ and $v = v_o$) having inside radius b and outside radius c.

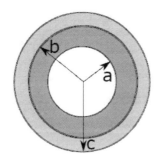

Figure 10-10

If the cylinders have been press fit together because there was a radial interference δ between the initial outer radius of the smaller cylinder and the initial inner radius of the larger cylinder, then an interface pressure p_I will develop at the contact boundary $r = b$. This interface pressure is

$$p_I = \frac{\delta}{b} \left[\frac{1}{E_i} \left(\frac{b^2 + a^2}{b^2 - a^2} - v_i \right) + \frac{1}{E_o} \left(\frac{c^2 + b^2}{c^2 - b^2} + v_o \right) \right]^{-1}$$

(*Note the negative exponent on the second term.*) This leaves the smaller cylinder effectively externally pressurized and the larger cylinder internally pressurized by p_I. If the two cylinders are made of the same material, then

$$p_I = \frac{E\delta}{2b^3} \frac{(b^2 - a^2)(c^2 - b^2)}{(c^2 - a^2)}$$

If there is a coefficient of friction between the two cylinders, the torque required to rotate the inner cylinder relative to the outer one is

$$Torque = 2\pi \mu b^2 p_I L$$

One technique for putting the cylinders together is to heat up the outer cylinder so that it's thermal expansion equals δ. Then the inner cylinder can be slid inside the outer one and the two cylinders allowed to cool. To accomplish this, we want

$$\frac{\delta}{b} = \alpha \Delta T$$

Therefore, we need to raise the temperature of the outer cylinder by

$$\Delta T = \frac{\delta}{\alpha b}$$

Example: A steel (with E=30x10⁶psi and v=0.3) collar with an outer diameter of 2in and a length of 2in is to be press fit onto a solid steel shaft with a diameter of 1in. If the coefficient of friction is μ=0.3, find the inner diameter of the collar that will allow the shaft to sustain a torque up to 10,000inlb without slipping.

For this case we use the nominal dimensions a=0, b=0.5in, and c=1in. The required interface pressure is

$$p_I = \frac{Torque}{2\pi\mu b^2 L} = \frac{10,000 inlb}{2\pi(0.3)(0.5in)^2(2in)} = 10,610\,psi$$

We now solve for the radial interference

$$\delta = \frac{2p_I b^3(c^2 - a^2)}{E(b^2 - a^2)(c^2 - b^2)} = \frac{2(10,610\,psi)(0.5in)^3[(1in)^2 - 0]}{30 \times 10^6\,psi[(0.5in)^2 - 0][(1in)^2 - (0.5in)^2]} = 0.00047in$$

The inside diameter of the collar needs to be

$$d_{collar} = 2b - 2\delta = 2(0.5in) - 2(0.00047in) = 0.99906in$$

10.6 Bolted Connections

10.6.1 Bolted joints under tension load

Bolts under tension load are effectively bars under uniaxial load where there is a normal stress equal to the tension force divided by the cross-section area. However, because of the thread, the load carrying area (stress area) A_t is less than the nominal area $A_d = \pi d^2/4$. The stress area for various size bolts is given in Table 10-1. The bolt stress is

$$\sigma = F_b/A_t$$

Table 10-1 Bolt sizes   Coarse Threads – UNC   Fine Threads – UNF

Nominal Size	Major Diameter (inch)	Threads/inch	Tensile Stress Area (square inches)	Threads/inch	Tensile Stress Area (square inches)
#1	0.073	64	0.003	72	0.003
#2	0.086	56	0.004	64	0.004
#3	0.099	48	0.005	56	0.005
#4	0.112	40	0.006	48	0.007
#5	0.125	40	0.008	44	0.008
#6	0.138	32	0.009	40	0.01
#8	0.164	32	0.014	36	0.015
#10	0.19	24	0.018	32	0.02
#12	0.216	24	0.024	28	0.026
1/4	0.25	20	0.032	28	0.036
5/16	0.313	18	0.052	24	0.058
3/8	0.375	16	0.078	24	0.088
7/16	0.438	14	0.106	20	0.119
1/2	0.5	13	0.142	20	0.16
9/16	0.563	12	0.182	18	0.203
5/8	0.625	11	0.226	18	0.256
3/4	0.75	10	0.334	16	0.373
7/8	0.875	9	0.462	14	0.509
1	1	8	0.606	14	0.663
1 1/8	1.125	7	0.763	12	0.856
1 1/4	1.25	7	0.969	12	1.073
1 3/8	1.375	6	1.155	12	1.315
1 1/2	1.5	6	1.405	12	1.581
1 3/4	1.75	5	1.900		

10.6.1.1 Initial tension and tightening torque

When a bolt is tightened, there is a relation between the initial tension force (preload) F_i and tightening torque T given by

$$T = 0.2 d F_i$$

where d=nominal bolt diameter.

If a bolt clamps two members together, and the members are under external load F_e as shown in Figure 10-11, the initial tension force needs to be large enough to keep the members pressed together.

Figure 10-11

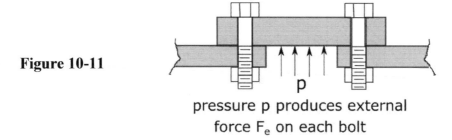

pressure p produces external
force F_e on each bolt

A more detailed view of the bolt is shown in Figure 10-12.

Figure 10-12

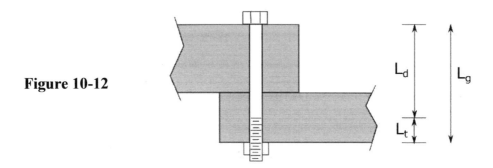

With F_e applied, the total load on the bolt is

$$F_b = CF_e + F_i \qquad \text{where} \quad C = \frac{k_b}{k_b + k_m}$$

and the total compressive load on the member is

$$F_m = (1 - C)F_e - F_i$$

where k_b is the effective stiffness of the bolt given as

$$k_b = \frac{A_d A_t E}{A_d L_t + A_t L_d}$$

where L_d=unthreaded length of the bolt
L_t=threaded length of the bolt
$L_G = L_d + L_t$=grip length

and k_m is the effective stiffness of the member. If all members under the compressive action of the bolt are of the same material, the effective member stiffness can be estimated from

$$k_m = 0.79 dE e^{0.63d/L_G}$$

The joint between the parts will open when F_m goes to zero. Therefore, the minimum initial load to prevent this is

$$F_{i-\min} = (1-C)F_e$$

Example: A 0.5in diameter coarse thread steel bolt in a steel pressure vessel carries an external force of 2,000lb. For this case we have L_d=2in and L_t=1in. Find the initial tension to keep the members from separating and the required tightening torque. Then find the stress in the bolt assuming the minimum initial tension is applied.

First, we calculate the stiffness of the bolt and the member.

$$k_b = \frac{A_d A_t E}{A_d L_t + A_t L_d} = \frac{[\pi(0.5in)^2/4](0.142in^2)30 \times 10^6 \, psi}{[\pi(0.5in)^2/4](1in)+(0.142in^2)(2in)} = 1.74 \times 10^6 \, lb/in$$

$$k_m = 0.79 dE e^{0.63d/L_G} = 0.79(0.5in)(30 \times 10^6 \, psi)e^{0.63(0.5in)/3in} = 13.16 \times 10^6 \, lb/in$$

$$C = \frac{k_b}{k_b + k_m} = \frac{1.74 \times 10^6 \, lb/in}{1.74 \times 10^6 \, lb/in + 13.16 \times 10^6 \, lb/in} = 0.117$$

The minimum initial tension is

$$F_{i-\min} = (1-C)F_e = (1-0.117)2,000lb = 1,766lb$$

The required torque is

$$T = 0.2dF_i = 0.2(0.5in)(1,766lb) = 177inlb$$

The total bolt load is

$$F_b = CF_e + F_i = 0.117(2,000lb) + 1,766lb = 2,000lb$$

The stress in the bolt is

$$\sigma = \frac{F_b}{A_t} = \frac{2,000lb}{0.142in^2} = 14,085psi$$

10.6.1.2 Power screws

If a nut is held from rotating and a screw is turned, the nut will move up or down on the screw. This device called a power screw can be used to raise or lower a weight as shown in Figure 10-13. The relation between the torque T on the screw and the force F delivered is as follows:

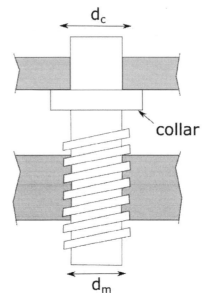

Figure 10-13

$$T_R = (\frac{\pi\mu d_m + \ell}{\pi d_m - \mu\ell})\frac{Fd_m}{2} + \frac{\mu_c d_c F}{2} \quad \text{to raise the weight}$$

$$T_L = (\frac{\pi\mu d_m - \ell}{\pi d_m + \mu\ell})\frac{Fd_m}{2} + \frac{\mu_c d_c F}{2} \quad \text{to lower the weight}$$

where d_m=mean bolt diameter=d-p/2

ℓ =lead=p=pitch=1/N=reciprocal of the number of threads per inch

μ=coefficient of friction of bolt threads

μ_c=coefficient of friction of collar

d_c=mean collar diameter=(d_{outer}+d_{inner})/2

Example: A power screw has a nominal diameter of 0.75in with 8 threads per inch and has a collar with a mean diameter of 2in. It is used to lift a weight of 500lb. If the coefficient of friction is 0.2 for the bolt thread and 0.3 for the collar, find the torque required to turn the screw.

First, we compute the lead.

$$\ell = p = 1/N = \frac{1}{8in^{-1}} = 0.125in$$

Next, we compute the mean diameter.

$$d_m = d - p/2 = 0.75in - 0.125in/2 = 0.6875in$$

The torque required to raise the weight is

$$T_R = (\frac{\pi\mu d_m + \ell}{\pi d_m - \mu\ell})\frac{Fd_m}{2} + \frac{\mu_c d_c F}{2} = \left(\frac{\pi(0.2)(0.6875in) + 0.125in}{\pi(0.6875in) - (0.2)(0.125in)}\right)\frac{(500lb)(0.6875in)}{2} + \frac{0.3(2in)(500lb)}{2}$$

$$T_R = 194.8inlb$$

10.6.2 Bolted joints under direct shear load

Consider a bolted joint under direct shear load as shown in Figure 10-14.

Figure 10-14

The shear stress in the bolt is

$$\tau = \frac{F}{A}$$

where A is the cross-section area of the bolt(s).

Example: Two plates are held together by three 0.5in diameter bolts as shown in Figure 10-15. Find the shear stress in each bolt if the load F=1,000lb.

Figure 10-15

Each bolt carries F/3. Therefore, the shear stress is

$$\tau = \frac{1000lb/3}{\pi(0.5in)^2/4} = 1,698psi$$

10.6.3 Bolted joints under eccentric shear load

When the load is applied eccentrically to the bolted members, the load is generally not shared equally by the bolts. For example, consider a beam with a bolted support as shown in Figure 10-16.

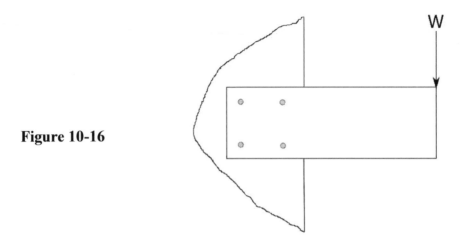

Figure 10-16

The reaction force and reaction moment must be carried by the bolts. Calculating the load carried by each bolt requires the following steps:

1. Find the center of mass of the bolts as shown in Figure 10-17.

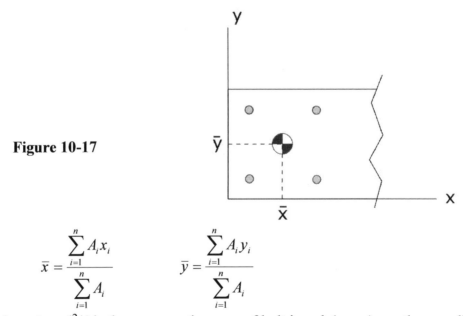

Figure 10-17

$$\bar{x} = \frac{\sum\limits_{i=1}^{n} A_i x_i}{\sum\limits_{i=1}^{n} A_i} \qquad \bar{y} = \frac{\sum\limits_{i=1}^{n} A_i y_i}{\sum\limits_{i=1}^{n} A_i}$$

where $A_i = \pi d^2/4$ is the cross-section area of bolt i, and (x_i, y_i) are the coordinates of bolt i.

2. Calculate the primary shear force $F_{pi}=W/n$ in each bolt which is the result of the reaction force W as shown in Figure 10-18.

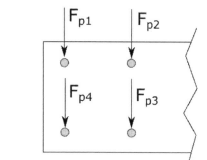

Figure 10-18

3. Calculate the reaction moment M=We as shown in Figure 10-19.

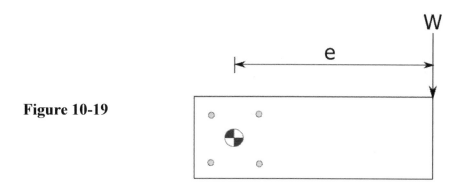

Figure 10-19

4. Calculate the secondary shear force F_{si} in each bolt which is the result of the reaction moment by assuming that the bolt force acts perpendicular to a radial line from the center of mass to the bolt as shown in Figure 10-20.

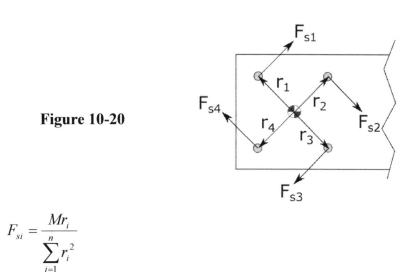

Figure 10-20

$$F_{si} = \frac{Mr_i}{\sum\limits_{i=1}^{n} r_i^2}$$

5. Find the resultant force F_{Ri} on each bolt from F_{pi} and F_{si} using vector algebra as shown in Figure 10-21.

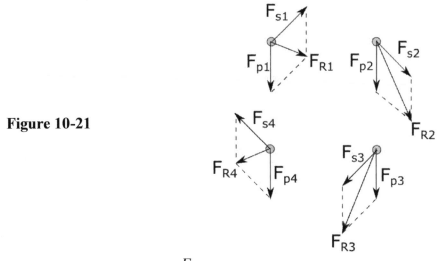

Figure 10-21

6. The maximum stress is $\tau_{max} = \dfrac{F_{R\,max}}{\pi d^2 / 4}$

Example: A rectangular steel plate carries a 2,00lb load and is held in place by two 1in diameter bolts as shown in Figure 10-22. Find the maximum bolt shear stress.

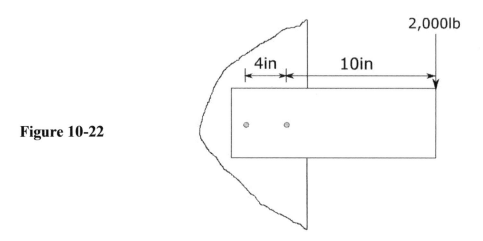

Figure 10-22

First, we observe that the center of mass of the bolts is midway between them. Then, we calculate W and M.

$$W = 2,000lb \quad \text{and} \quad M = 2,000lb(12in) = 24,000inlb$$

The primary shear force is

$$F_{pi} = W / N = 2,000lb / 2 = 1,000lb$$

The center of mass is at the center of the bolt pattern. The secondary shear force in each bolt is

$$F_{si} = \frac{Mr}{2r^2} = \frac{M}{2r} = \frac{24,000inlb}{2(2in)} = 6,000lb$$

Now we superpose the primary and secondary shear forces as shown in Figure 10-23.

Figure 10-23

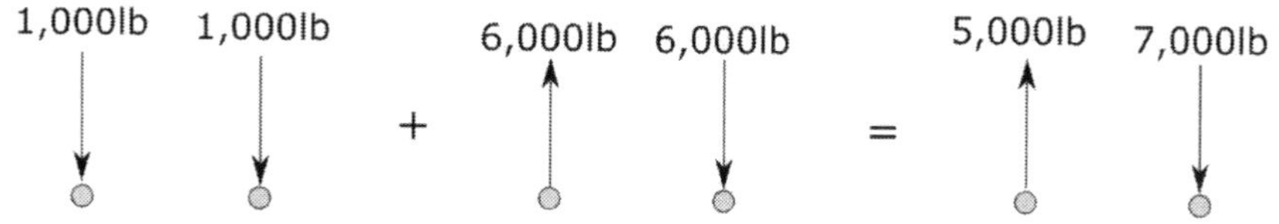

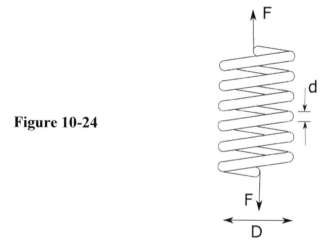

The largest shear force is F_{Rmax}=7,000lb. Therefore, the maximum shear stress is

$$\tau_{max} = \frac{F_{R max}}{\pi d^2 / 4} = \frac{7,000lb}{\pi (1in)^2 / 4} = 8,913psi$$

10.7 Springs

10.7.1 Helical linear springs

Consider a helical linear spring under an axial load F as shown in Figure 10-24.

Figure 10-24

The maximum shear stress is a combination of direct shear and torsion

$$\tau = K \frac{8FD}{\pi d^3}$$

where $K = \dfrac{2C+1}{2C}$ and C=D/d, D=mean spring diameter, and d=wire diameter

The shear strength is normally taken to be half the yield strength in tension. Therefore, failure occurs when

$$\tau = S_y / 2$$

The spring constant relating force to displacement is

$$k = \frac{Gd^4}{8ND^3}$$

where G is the shear modulus of the material and N is the number of active coils.

Example: A spring with a mean diameter of 4in is made of 0.25in diameter wire with G=12x10^{6}psi. If the spring has 10 coils, find the shear stress and the stretch of the spring under a load of 10lb.

First, we calculate C=D/d=4in/0.25in=16

Now, we find K.

$$K = \frac{2C+1}{2C} = \frac{2(16)+1}{2(16)} = 1.03$$

The shear stress is

$$\tau = K\frac{8FD}{\pi d^3} = 1.03\frac{8(10lb)(4in)}{\pi(0.25in)^3} = 6,715\,psi$$

The spring stiffness is

$$k = \frac{Gd^4}{8ND^3} = \frac{(12\times10^6)(0.25in)^4}{8(10)(4in)^3} = 9.16lb/in$$

For a load F=10lb, the stretch is

$$\delta = \frac{F}{k} = \frac{10lb}{9.16lb/in} = 1.09in$$

10.7.2 Helical torsion springs

Consider a helical torsion spring under load F with moment arm r as shown in Figure 10-25.

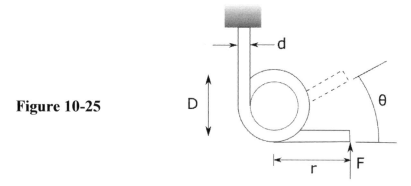

Figure 10-25

The normal stress due to bending is

$$\sigma = K_t \frac{32Fr}{\pi d^3}$$

where

$$K_t = \frac{4C^2 - C - 1}{4C(C - 1)} \qquad \text{where C=D/d}$$

The spring constant k relating rotation θ to moment Fr is

$$k = \frac{Fr}{\theta} = \frac{Ed^4}{64ND}$$

Example: A torsion spring with D=1in, d=0.125in, and r=1.5in is subjected to a force of 2lb. Find the bending normal stress.

First, we calculate K_t with C=D/d=1in/0.125in=8.

$$K_t = \frac{4C^2 - C - 1}{4C(C - 1)} = \frac{4(8)^2 - 8 - 1}{4(8)(8 - 1)} = 1.10$$

$$\sigma = K_t \frac{32Fr}{\pi d^3} = 1.10 \frac{32(2lb)(1.5in)}{\pi(0.125in)^3} = 17,210 psi$$

10.8 Gears

10.8.1 Spur gears

The pitch circle is the theoretical circle used for angular velocity calculations, etc. The pitch circles of two mating gears are tangent to each other.

> d=diameter of pitch circle
> N=number of teeth
> P=diametral pitch=N/d

The smaller of two mating gears is called the pinion.

10.8.2 Gear trains

From kinematics of rigid bodies, we know that the velocity of the edge of gear i is $v_i=\omega_i r_i$. The contact points of two mating gears have equal velocities, i.e.,

$$\omega_1 r_1 = \omega_2 r_2$$

Therefore, noting that the radius is proportional to the number of teeth, we get

$$\omega_2 = \frac{r_1}{r_2}\omega_1 = \frac{N_1}{N_2}\omega_1$$

Now consider a train of interconnecting gears each rotating about it center as shown in Figure 10-26.

Figure 10-26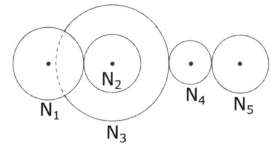

Matching the edge velocities, we get the following relations.

$$\omega_1 N_1 = \omega_2 N_2, \quad \omega_3 = \omega_2, \quad \omega_3 N_3 = \omega_4 N_4, \quad \omega_4 N_4 = \omega_5 N_5$$

If we combine these equations, we get

$$\omega_5 = \frac{N_1 N_3 N_4}{N_2 N_4 N_5}\omega_1$$

For the general case we have

$$\omega_{out} = m_v \omega_{in}$$

where m_v is the velocity ratio given by

$$m_v = \frac{product \cdot of \cdot number \cdot of \cdot teeth \cdot on \cdot driver \cdot gears}{product \cdot of \cdot number \cdot of \cdot teeth \cdot on \cdot driven \cdot gears}$$

The relationship for torque follows the inverse of that for angular velocity; i.e.,

$$T_{out} = T_{in} / m_v$$

10.8.3 Gear forces and tooth stress

The force on a gear tooth W can be resolved into a radial component W_r and a tangential component W_t as shown in Figure 10-27.

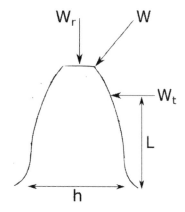

Figure 10-27

The tangential force W_t is related to the transmitted torque T through

$$W_t = \frac{T}{r}$$

Power H is related to torque through

$$H = T\omega$$

where ω is angular velocity. From this we can write

$$W_t = \frac{H}{\omega r}$$

The tangential force causes bending stress as in a cantilever beam (see the Figure 10-27 above)

$$\sigma = \frac{6W_t L}{bh^2}$$

where b is the depth of the gear.

Example: A motor transmits 20hp at 3,000rpm. Find the tangential force on a gear with r=4in.

First, let's calculate torque

$$T = \frac{H}{\omega} = \frac{20hp}{3000rev/min}\left(\frac{550\,ftlb/s}{1hp}\right)\left(\frac{12in}{1ft}\right)\left(\frac{1rev}{2\pi rad}\right)\left(\frac{60s}{1min}\right) = 420inlb$$

The tangential force is

$$W_t = \frac{T}{r} = \frac{(420inlb)}{4in} = 105lb$$

10.9 Shafts

A shaft, as shown in Figure 10-28, will typically be subjected to a combination of bending moment M and torsional moment T.

Figure 10-28

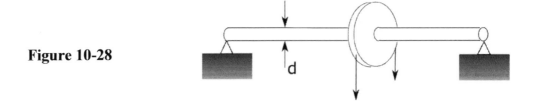

This produces normal stress and shear stress. The maximum shear stress from this combination is

$$\tau_{max} = \sqrt{(\frac{\sigma}{2})^2 + \tau^2} = \frac{16}{\pi d^3}\sqrt{M^2 + T^2}$$

The yield stress in shear is typically taken to be one half the yield stress in tension; i.e.,

$$\tau_y = S_y/2$$

Then, failure occurs when $\tau_{max} = S_y/2$

or $\quad \dfrac{16}{\pi d^3}\sqrt{M^2 + T^2} = S_y/2$

Example: Find the maximum shear stress in the shaft shown in Figure 10-29.

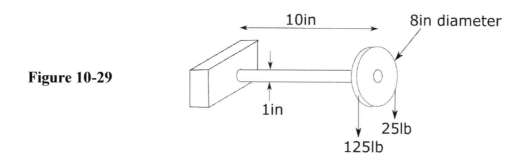

Figure 10-29

The shaft effectively acts as a cantilever beam with an end force of 150lb and end torque of T=4in(125lb-25lb)=400inlb as shown in Figure 10-30.

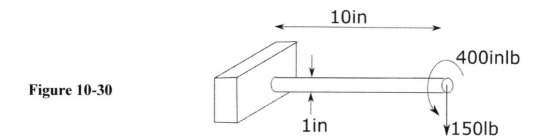

Figure 10-30

The maximum moment occurs at the left end of the shaft and is equal to

$$M = 150lb(10in) = 1,500inlb$$

The maximum shear stress is

$$\tau_{max} = \frac{16}{\pi d^3}\sqrt{M^2 + T^2} = \frac{16}{\pi(1in)^3}\sqrt{(1,500inlb)^2 + (400inlb)^2} = 7,900\,psi$$

10.10 Bearings

If two groups of identical bearings are tested under different loads F_1 and F_2, then their respective lives are related by

$$\frac{N_1}{N_2} = \left[\frac{F_2}{F_1}\right]^k$$ where k=3 for ball bearings and k=10/3 for roller bearings.

and N is life in number of revolutions. The life in hours L is related to N by

$$N = 60\omega L$$

where ω is speed in rpm, or

$$\frac{\omega_1 L_1}{\omega_2 L_2} = \left(\frac{F_2}{F_1}\right)^k$$

Example: A roller bearing is rated to be capable of sustaining 1 million revolutions under a load of 200lb. If the bearing is used in a piece of rotating machinery operating at 2000rpm, how many hours will the bearing last under a load of 300 pounds?

$$\frac{N_1}{60\omega L_2} = \left(\frac{F_2}{F_1}\right)^k \rightarrow L_2 = \frac{N_1}{60\omega}\left(\frac{F_1}{F_2}\right)^k = \frac{1\times10^6}{60(2000)}\left(\frac{200lb}{300lb}\right)^{10/3} = 2.16hr$$

11. THERMODYNAMICS

11.1 Properties of Fluids

Thermodynamics typically deals with substances in their liquid and/or gas form. The properties of interest are mass m, volume V, pressure p, and temperature T. Mass and volume can be combined together as volume per unit mass or specific volume v defined as

$$v = V / m$$

which is the inverse of density (In the US Customary System, the most commonly used unit for mass is the pound mass, lbm=1/32.2 slug). There are four temperature scales in common use: two relative scales (Celcius and Farenheit) and two absolute scales (Kelvin and Rankine). These are related as follows:

$$T(°R)=T(°F)+460°$$
$$T(°K)=T(°C)+273°$$
$$T(°R)=1.8T(°K)$$
$$T(°F)=1.8T(°C)+32°$$

Water freezes at 32°F (0°C) and boils at 212°F (100°C).

We can raise the temperature of a substance by heating it. The increase in internal energy (molecular vibration, etc) can be represented by the property u (internal energy per unit mass). The increase in internal energy is linked to the increase in temperature by the specific heat for constant volume C_v; i.e.,

$$\Delta u = C_v \Delta T$$

Another property of interest is the specific enthalpy h defined as

$$h = u + pv$$

which is a combination of internal energy and flow work (pv). The increase in enthalpy is linked to the increase in temperature by the specific heat for constant pressure C_p; i.e.,

$$\Delta h = C_p \Delta T$$

Values for C_v and C_p can be found in Table 11-1 . These values tend to depend on temperature.

Table 11-1

	R kJ/kg°K	C$_p$ kJ/kg°K	C$_v$ kJ/kg°K	k
Air	0.2870	1.005	0.718	1.4
Helium	2.077	5.1926	3.1156	1.667
Nitrogen	0.2968	1.039	0.743	1.4
Oxygen	0.2598	0.918	0.658	1.395

11.1.1 Ideal gas

Many gases at low density follow the ideal gas law

$$pv = RT$$

where $R = \bar{R}/M$. $\bar{R}$ is the universal gas constant 8.314kJ/kmol°K, and M is the mass of one mole of a substance in grams. The values of R for several gases are listed in Table 11-1. Using the ideal gas equation requires that T be expressed in absolute temperature (°K or °R). From this equation it is clear that only two of the quantities p, v, and T are independent. For many gases, we can use tables (see Appendix A) to determine p, v, u, and h for a value of temperature.

11.1.2 Water

For systems involving water, we must deal with water in both its liquid form and vapor form (vapor means gas). Again, the properties p, v, and T are interrelated. Let's consider heating liquid water at constant atmospheric pressure to convert it to steam. This process involves several steps as indicated in the five states listed below.

    1. Liquid below the boiling point (supercooled liquid)
    2. Liquid at the boiling point (saturated liquid)
    3. A mixture of liquid and vapor at the boiling point (saturated liquid/vapor mixture)
    4. Vapor at the boiling point (saturated vapor)
    5. Vapor above the boiling point (superheated vapor)

As heat is added, the temperature of the liquid water rises until it reaches the boiling point 212°F (100°C) at atmospheric pressure. At this point the water remains at a constant 212°F (100°C) until all the liquid has been converted to vapor even though heat is being continuously delivered. After this point, the vapor increases in temperature as heat is added. This process is illustrated in a plot of T versus v as shown in Figure 11-1.

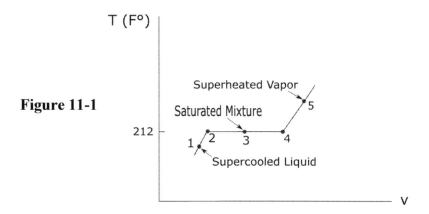

Figure 11-1

Between points 1 and 2, the liquid water rises in temperature as heat is added. During this step the water is called a supercooled liquid or compressed liquid. When it reaches point 2 and is about to vaporize, it is called a saturated liquid. Between points 2 and 4 the water gradually changes from liquid phase to vapor phase. At some intermediate point (point 3), there is a mixture of liquid and vapor. In this region we define the quality x as the ratio of the mass of the vapor (gas) to the total mass; i.e.,

$$x = \frac{m_{gas}}{m_{total}} = \frac{m_g}{m_T}$$

We can divide the specific volume v into specific volume of liquid (flussigkeit in German) v_f and specific volume of vapor (gas in German) v_g so that

$$v = (1-x)v_f + xv_g$$
$$\text{or} \quad v = v_f + xv_{fg} \quad \text{where} \quad v_{fg} = v_g - v_f$$

At point 4 when all the liquid has been converted to vapor, we refer to this state as saturated vapor. Between points 4 and 5, the vapor increases in temperature as heat is added. We refer to this state as superheated vapor.

If we made a T versus v plot for water at a higher pressure, we would get a similar curve, but it would be shifted upward as shown in Figure 11-2.

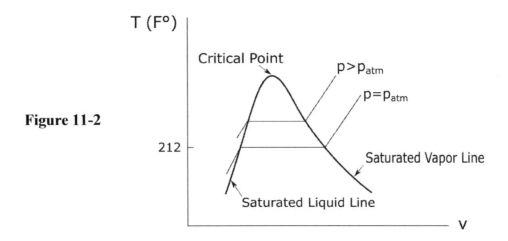

Figure 11-2

Suppose we repeat this T versus v curve for many different values of p. We could pass a line through the points where the liquid begins to vaporize. This is the saturated liquid line on the plot above. We could also pass a line through the points were the vapor begins to condense back to liquid. This would be the saturated vapor line on the plot above. The point where these two lines meet at the top is called the critical point. Properties of water in various states can be found in steam tables (see Appendix A).

11.2 First Law of Thermodynamics

The first law of thermodynamics is a conservation of energy statement. We can apply this to a closed or open system. In a closed system, we deal with a fixed mass of material. In an open system, we deal with a fixed volume (control volume) which material passes through. These two cases are illustrated in Figure 11-3.

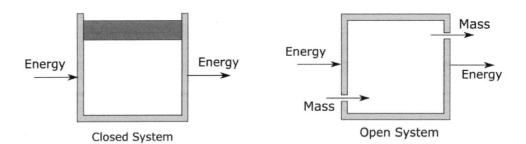

Figure 11-3

11.2.1 Analysis of a closed system

For a closed system the first law states the following

(Total Energy)$_{in}$-(Total Energy)$_{out}$=Change in Energy

Energy can be in the form of heat Q or work W. Then

$$(Q_i + W_i) - (Q_o + W_o) = \Delta E$$

Example: Heating of a gas in a cylinder

A piston sits on a cylinder containing air and maintains a constant pressure of 1000kPa. Initially, the air is at 40°C with a volume of $0.1m^3$. If 20kJ of heat is added to the air, find its final volume and temperature.

We can treat the air as an ideal gas and use the ideal gas law to find the mass; i.e.,

$$pv = RT \rightarrow p\frac{V}{m} = RT$$

$$m = \frac{p_1 V_1}{RT_1} = \frac{(1000 \times 10^3 \, Newton/m^2)(0.1m^3)}{(0.287kJ/kg°K)(313°K)}\left(\frac{1kJ}{1000 \, Newtonm}\right) = 1.11kg$$

Writing the ideal gas law as $R = \dfrac{pV}{mT}$, we can write

$$\frac{p_1 V_1}{mT_1} = \frac{p_2 V_2}{mT_2}$$

Because $p_2 = p_1$, we get

$$V_2 = V_1 \frac{T_2}{T_1}$$

As the gas expands under constant pressure, we have work output as

$$W_o = \int p \, dV = p_1 \int dV = p_1(V_2 - V_1) = p_1\left(\frac{V_1 T_2}{T_1} - V_1\right)$$

$$W_o = p_1 V_1\left(\frac{T_2}{T_1} - 1\right) = (1000 \times 10^3 \, Newt/m^2)(0.1m^3)\left(\frac{T_2}{313°K} - 1\right) = 100kJ\left(\frac{T_2}{313°R} - 1\right)$$

The change in energy in the gas can be written as

$$\Delta E = m(u_2 - u_1) = mC_v(T_2 - T_1)$$

$$\Delta E = 1.11kg(0.718kJ/kg°K)(T_2 - 313°K) = 249kJ\left(\frac{T_2}{313°R} - 1\right)$$

Applying the first law gives

$$(Q_i + W_i) - (Q_o + W_o) = \Delta E$$

$$[20kJ + 0] - \left[0 + 100kJ \left(\frac{T_2}{313°R} - 1 \right) \right] = 249kJ \left(\frac{T_2}{313°R} - 1 \right)$$

Solving gives $T_2 = 331°K = 58°C$.

The final volume is

$$V_2 = V_1 \frac{T_2}{T_1} = 0.1m^3 \left(\frac{331°K}{313°K} \right) = 0.106m^3$$

Example: Cooling of steam
A constant volume vessel contains steam at a pressure of 100psia and at a temperature of 500°F. Find the temperature to which the steam must be cooled to reach a quality of 50%
.

From the superheated steam table A-3 with p_1=100pisa and T_1=500°F, v_1=5.587ft³/lbm. Since the volume does not change, v_2=v_1=5.587ft³/lbm at the final state. Also,

v_2=(1-x)v_{f2}+xv_{g2}

We expect v_{f2} to be negligible compared to v_{g2}. Then

5.587ft³/lbm=0.5v_{g2} → v_{g2}=11.17ft³/lbm

Using this value of v_{g2} and the saturated steam table A-1 in Appendix A, we get T_2=264°F.

11.2.2 Analysis of an open system with steady flow

For this type of system we have a control volume where the mass flowing in is equal to the mass flowing out. In this case we have both a mass balance and energy balance. The mass flow rate can be expressed as

$$\dot{m} = \rho \upsilon A = \frac{\upsilon}{v} A$$

where υ is the flow velocity, and A is the cross section area. The mass balance requires

$$\sum \dot{m}_i = \sum \dot{m}_o$$

The energy balance requires

$$\dot{Q}_i + \dot{W}_i + \sum \dot{m}_i (h_i + \frac{\upsilon_i^2}{2} + gz_i) = \dot{Q}_o + \dot{W}_o + \sum \dot{m}_o (h_o + \frac{\upsilon_o^2}{2} + gz_o)$$

where $\dot{Q}$ is the heat rate, $\dot{W}$ is the work rate (power), h=u+pv is the specific enthalpy, υ is the flow velocity (kinetic energy effect), and z is the elevation (potential energy effect). Many thermodynamic devices operate under steady flow conditions. Several cases are described below.

11.2.3 Mixing of fluids

When two or more streams of fluids are mixed together to form a single stream, we need to apply both conservation of mass and conservation of energy. Typically, we can ignore changes in kinetic and potential energies as well as work and heat entering or leaving the system. Then

$$\sum \dot{m}_i = \dot{m}_o \quad \text{and} \quad \sum \dot{m}_i h_i = \dot{m}_o h_o$$

Example: Mixing of two streams of liquid water
Suppose 150°F water flowing through a pipe at 2lbm/s is mixed with 70°F water flowing through another pipe at 1lbm/s, all at atmospheric pressure. Find the temperature of the output flow.

From conservation of mass

$$\dot{m}_o = \dot{m}_1 + \dot{m}_2 = 2lbm/s + 1lbm/s = 3lbm/s$$

From the first law

$$\dot{m}_1 h_1 + \dot{m}_2 h_2 = \dot{m}_o h_o$$

The enthalpy of liquid water is approximately the same as the enthalpy of saturated water. From the steam table A-1

At 150°F, h_1=118.07BTU/lbm; and at 70°F, h_2=38.1BTU/lbm

Then
(2lbm/s)(118.07BTU/lbm)+(1lbm/s)(38.1BTU/lbm)=(3lbm/s)h_o

Solving for h_o gives h_o=91.41BTU/lbm. Using this value and steam table A-1 gives T_2=123.4°F.

11.2.4 Nozzles and diffusers

A nozzle is a device that increases the velocity of a fluid with a corresponding lowering of pressure as fluid flows through a large area inlet to a smaller area outlet. A diffuser is a device that decreases the velocity of a fluid with a corresponding increase in pressure as fluid flows through a small area inlet to a larger area outlet.

For this case, $\dot{Q}_i = 0, \dot{W}_i = 0, \dot{Q}_o = 0, \dot{W}_o = 0, z_i = z_o, \dot{m}_i = \dot{m}_o$. Therefore, the first law gives

$$h_i + \upsilon_i^2/2 = h_0 + \upsilon_o^2/2$$

Example: Steam flow through a nozzle
Steam flows at a steady rate through a nozzle with an inlet velocity 200ft/s and an outlet velocity of 1000ft/s. The inlet pressure and temperature are p_i=200psia and T_i=600°F, respectively. The outlet pressure is p_o=100psia. Find the outlet temperature if heat losses are negligible.

The first law gives

$$h_0 = h_i + \upsilon_i^2/2 - \upsilon_o^2/2$$

From the superheated steam table A-3 with p_i=200psia and T_i=600°F, h_i=1323.2BTU/lbm. Then

$$h_o = 1323.2\,BTU/lbm + \left(\frac{(200\,ft/s)^2 - (1000\,ft/s)^2}{2}\right)\left(\frac{1BTU/lbm}{25{,}037\,ft^2/s^2}\right) = 1304\,BTU/lbm$$

From the superheated steam tables with p_o=100psia and h_o=1304BTU/lbm, we get T_o=548°F.

11.2.5 Turbines and compressors (pumps, fans)

A turbine is a device where a fluid enters at a high pressure and leaves at a lower pressure as the turbine produces an output of work. For this case, we have $\dot{Q}_i = 0, \dot{W}_i = 0, \dot{Q}_o = 0$ and gz is negligible for a gas. The first law gives

$$\dot{m}_i(h_i + \upsilon_i^2/2) = \dot{m}_o(h_0 + \upsilon_o^2/2) + \dot{W}_o$$

A compressor (pump, fan) is a device where a fluid enters in at a low pressure and leaves at a higher pressure as work is input to the compressor. For this case the first law gives

$$\dot{m}_i(h_i + \upsilon_i^2/2) + \dot{W}_i = \dot{m}_o(h_0 + \upsilon_o^2/2)$$

Example: Steam Turbine

Superheated steam enters a turbine at p_i=300psia, T_i=700°F, and $\upsilon_1 = 200 ft/s$. A saturated mixture with 80% quality exits at T_o=150°F and $\upsilon_2 = 100 ft/s$. If the mass flow rate is 10lbm/s and heat losses are negligible, find the power output of the turbine.

From the superheated steam tables with p_i=300psia and T_i=700°F, h_i=1369.5BTU/lbm.

From the saturated steam tables with T_o=150°F, h_{fo}=118.07BTU/lbm and h_{fgo}=1008.4BTU/lbm.

Then $h_o = h_{fo} + x h_{fgo} = 118.07 BTU/lbm + (0.8)(1008.4 BTU/lbm) = 924.4 BTU/lbm$
The first law gives

$$\dot{W}_o = \dot{m}_i(h_i + \upsilon_i^2/2) - \dot{m}_o(h_0 + \upsilon_o^2/2)$$

$$\dot{W}_o = 10 lbm/s \left[1369.5 BTU/lbm - 924.4 BTU/lbm + \left(\frac{(200 ft/s)^2}{2} - \frac{(100 ft/s)^2}{2} \right) \left(\frac{1BTU/lbm}{25,037 ft^2/s_2} \right) \right]$$

$$\dot{W}_o = 4450 BTU/s \left(\frac{3600s}{1hr} \right) = 16.0x10^6 BTU/hr \left(\frac{1Watt}{3.412 BTU/hr} \right) = 4.69 MW$$

If the effects of velocity change are ignored, the power calculation changes only by a fraction of a percent. A good approximation of the power output of a turbine is

$$\dot{W}_o = \dot{m}(h_i - h_0)$$

For the case of a compressor, the input power is

$$\dot{W}_i = \dot{m}(h_o - h_i)$$

If the fluid is a liquid rather than a gas, compressibility effects can be ignored, and the change in u in h=u+pv is negligible. In this case the input power required by a pump is

$$\dot{W}_i = \dot{m}v(p_o - p_i)$$

11.3 Second Law of Thermodynamics

Many thermodynamic systems contain a fluid that acts in a cycle. For example, consider the steam power plant system shown in Figure 11-4.

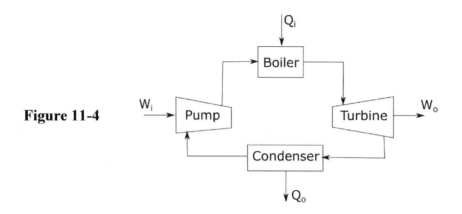

Figure 11-4

Liquid water passes through a pump where work is input to increase the fluid pressure. It then passes through a boiler where it is heated to form high-temperature steam. Next, it passes through a turbine that outputs work. It leaves the turbine as low temperature steam or as a mixture of liquid water and steam. Next, it passes through a condenser where heat is removed as all of the steam is converted to liquid water. Then, this cycle repeats. The net output of work is

$$W_{net} = W_o - W_i$$

From the first law of thermodynamics, we have

$$W_{net} = Q_i - Q_o$$

We define the thermal efficiency η_{th} of this cycle as

$$\eta_{th} = \frac{W_{net}}{Q_i} = \frac{Q_i - Q_o}{Q_i} = 1 - \frac{Q_o}{Q_i}$$

The input heat is delivered from a high temperature source, so let us rename $Q_i = Q_H$. The output heat is removed by a low temperature source, so let us rename $Q_o = Q_L$. Then

$$\eta_{th} = 1 - \frac{Q_L}{Q_H}$$

The Kelvin-Planck statement of the second law of thermodynamics says "It is not possible for any device that operates in a cycle to produce a net amount of work while receiving heat from a single source." This means that the cycle for the steam power plant will not function without the condenser.

For an idealized system called a Carnot cycle (where each step is reversible), it can be shown that

$$\frac{Q_L}{Q_H} = \frac{T_L}{T_H}$$

For this idealized system, the efficiency is

$$\eta_{carnot} = 1 - \frac{T_L}{T_H}$$

This is the maximum efficiency possible for a given system. In reality, most systems have efficiencies less than the Carnot efficiency.

11.3.1 Entropy

For application of the second law, it is useful to define the property entropy as

$$\Delta s = \int \frac{dQ}{mT}$$

For example, consider the change in entropy in converting saturated liquid water at 212°F at constant atmospheric pressure to saturated vapor by heating it. Since the temperature remains constant at 212°F(=672°R) during this process, we have

$$\Delta s = \frac{1}{mT} \int dQ = \frac{1}{mT}(Q_2 - Q_1)$$

To determine Q_2-Q_1 , we will apply the first law to a closed system.

$$Q_2 - Q_1 = Q_i$$
$$Q_o = 0$$
$$W_i = 0$$

When the water is converted from liquid to vapor at constant pressure, its volume will expand and work will be output; i.e.,

$$W_o = \int p dV = p \int dV = pV_2 - pV_1$$

The change in energy of the fluid is $\Delta E = u_2 - u_1$. Therefore, applying the first law

$$(Q_i + W_i) - (Q_o + W_o) = \Delta E$$
$$Q_2 - Q_1 - (pV_2 - pV_1) = m(u_2 - u_1)$$
$$\frac{Q_2 - Q_1}{m} = (u_2 + \frac{pV_2}{m}) - (u_1 + \frac{pV_1}{m}) = h_2 - h_1$$

Then $\quad \Delta s = \dfrac{h_2 - h_1}{T}$

From steam table A-2 h_2=1150.5BTU/lbm and h_1=180.15BTU/lbm.

Then $\quad \Delta s = \dfrac{1150.5 BTU/lbm - 180.15 BTU/lbm}{672°R} = 1.444 BTU/lbm°R$

The change in entropy could also have been determined directly from the steam tables as

$$\Delta s = s_2 - s_1 = 1.7557 BTU/lbm°R - 0.3121 BTU/lbm°R = 1.444 BTU/lbm°R$$

If we have a saturated mixture of liquid and vapor, then

$$s = s_f + x s_{fg}$$

In general, we can make a plot of T versus s for water just like we did for T versus v as shown in Figure 11-5.

Figure 11-5

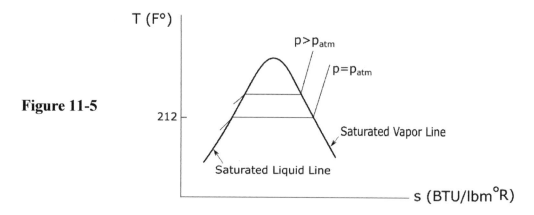

11.3.2 Isentropic process

Processes where the entropy remains constant are called isentropic. This happens when a process is reversible and adiabatic (no heat transfer). Ideally, turbines and compressors (pumps) would function as an isentropic process. Their actual performance is generally not isentropic. The closer they get to functioning isentropically, the better. We will quantify this comparison with the term isentropic efficiency.

11.3.3 Turbine efficiency

For a turbine the isentropic efficiency is

$$\eta_T = \frac{h_1 - h_2}{h_1 - h_{2s}}$$

where h_2 is the actual exit enthalpy, and h_{2s} is the isentropic exit enthalpy.

Example: Steam Turbine
Steam flows through a turbine with an entrance pressure p_1=300psia and temperature T_1=600°F and exits as a saturated vapor at T_2= 190°F. Find the isentropic efficiency.

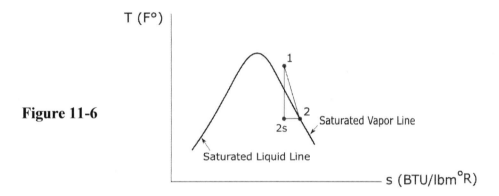

Figure 11-6

The process is shown in Figure 11-6 where the idealized isentropic process is represented by path 1-2s, and the actual process that does not have constant entropy is represented by path 1-2.

For state 1 (p_1=300psia T_1=600°F), the steam tables give h_1=1315.7BTU/lbm and s_1=1.6282BTU/lbm°R.

For idealized state 2s with s_2=s_1=1.6282BTU/lbm°R, we have a mixture of saturated liquid and vapor at T_2=190°F, and the steam tables give

s_{2s}=s_{2sf}+xs_{2sfg}

1.6282BTU/lbm°R =0.2789BTU/lbm°R +x(1.5153BTU/lbm°R)

Solving gives the quality at state 2s as x=0.890.

Therefore, the enthalpy at idealized state 2s can be determined from the steam tables as

h_{2s}=h_{2sf}+xh_{2sfg}
h_{2s}=158.18BTU/lbm+(0.890)(984.42BTU/lbm)=1034.3BTU/lbm

For actual state 2 that consists of a saturated vapor at 190°F, the steam tables give h_2=1142.6BTU/lbm.

The isentropic efficiency is then

$$\eta_R = \frac{h_1 - h_2}{h_1 - h_{2s}}$$

$$\eta_R = \frac{1315.7 BTU / lbm - 1142.6 BTU / lbm}{1315.7 BTU / lbm - 1034.3 BTU / lbm} = 62\%$$

11.3.4 Entropy in ideal gases

The entropy change for an ideal gas can be expressed as

$$s_2 - s_1 = \int C_V \frac{dT}{T} + R \ln \frac{v_2}{v_1}$$

$$s_2 - s_1 = \int C_P \frac{dT}{T} - R \ln \frac{p_2}{p_1}$$

If the specific heat values are relatively constant, then these formulas simplify to

$$s_2 - s_1 = C_v \ln \frac{T_2}{T_1} + R \ln \frac{v_2}{v_1}$$

$$s_2 - s_1 = C_p \ln \frac{T_2}{T_1} - R \ln \frac{p_2}{p_1}$$

If the process is isentropic ($S_2 - S_1 = 0$), then

$$\frac{T_2}{T_1} = (\frac{p_2}{p_1})^{(k-1)/k} = (\frac{v_1}{v_2})^{k-1}$$

where $k = C_p / C_v$.

Example: Air is compressed isentropically from an initial state of 10kPa and 15°C to a final temperature of 160°C. Find the final pressure.

$$\frac{T_2}{T_1} = (\frac{p_2}{p_1})^{(k-1)/k} \qquad \text{or} \qquad p_2 = p_1 (\frac{T_2}{T_1})^{k/(k-1)}$$

For air $k = 1.4$

Then $p_2 = 10 kPa \left(\frac{433°K}{288°K} \right)^{1.4/(1.4-1)} = 41.7 kPa$

11.3.5 Compressor (pump) efficiency

The isentropic efficiency of a compressor (pump) is

$$\eta_c = \frac{h_{2s} - h_1}{h_2 - h_1}$$

Example: Air compressor
Air at atmospheric pressure (p_1=14.7psia) and temperature T_1=80°F is compressed to pressure p_2=100pisa and temperature T_2=520°F. Find the isentropic efficiency of the compressor.

We will ignore the slight effect of temperature on values of specific heat. For air at 80°F, we have

C_p=0.240BTU/lbm°R, C_v=0.171BTU/lbm°R, k=1.4

For ideal gas

$$\frac{T_2}{T_1} = (\frac{p_2}{p_1})^{(k-1)/k}$$

Therefore, the theoretical exit temperature for an isentropic process is

$$T_2 = 540°R(\frac{100\,psia}{14.7\,psia})^{(1.4-1)/1.4} = 934°R = 474°F$$

The theoretical change in enthalpy is

$$\Delta h = C_P(T_2 - T_1) = 0.240BTU\,/\,lbm°R(934°R - 540°R) = 94.56BTU\,/\,lbm$$

From air table A-4 in Appendix A at T_1 =540°R, h_1=128.9BTU/lbm, and the theoretical enthalpy at state 2 is

$$h_{2s} = h_1 + \Delta h = 128.9BTU\,/\,lbm + 94.56BTU\,/\,lbm = 223.5BTU\,/\,lbm$$

From air table A-4 for the actual exit temperature T_2=980°R, we get

h_2=236.0BTU/lbm

Therefore, the isentropic efficiency is

$$\eta_c = \frac{h_{2s} - h_1}{h_2 - h_1}$$

$$\eta_c = \frac{223.5 BTU/lbm - 128.9 BTU/lbm}{236.0 BTU/lbm - 128.9 BTU/lbm} = 88.3\%$$

11.3.6 Refrigeration systems

The efficiency of a refrigeration system (including air conditioners) is given as the coefficient of performance (COP); i.e.,

$$COP_R = \frac{1}{Q_H/Q_L - 1}$$

For the idealized case of a Carnot cycle

$$COP_{carnot} = \frac{1}{T_H/T_L - 1}$$

A common refrigeration system is based on a vapor compression refrigeration cycle. It consists of four processes as shown in Figure 11-7. Each process is also depicted in a T versus s diagram and a p versus h diagram in Figure 11-8.

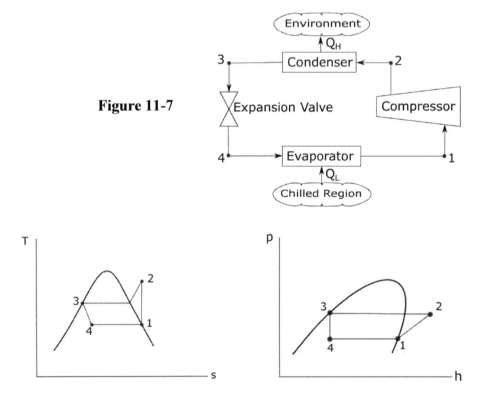

Figure 11-7

Figure 11-8

Compressor (Stage 1-2): Refrigerant enters the compressor as a saturated vapor and is compressed to temperature T_2 and pressure p_2 while s remains constant.

Condenser (Stage 2-3): Refrigerant enters the condenser as a superheated vapor and is condensed to a saturated liquid at temperature T_3 while p remains constant. Heat is transferred out of the refrigerant to the outside environment.

Expansion Valve (Stage 3-4): Refrigerant is throttled to a mixture of liquid and vapor at temperature T_4 while h remains constant.

Evaporator (Stage 4-1): Refrigerant completely evaporates to a saturated vapor at temperature $T_4 = T_1$ while p remains constant. Heat is absorbed from the refrigerator space into the refrigerant.

The coefficient of performance is

$$COP_R = \frac{\dot{Q}_L}{\dot{W}_{in}} = \frac{h_1 - h_4}{h_2 - h_1} \quad \text{and} \quad COP_{RCarnot} = \frac{1}{T_H / T_L - 1}$$

$\dot{Q}_L$ is often expressed in tons of refrigeration where

    1 ton=12000BTU/hr=3.516kW

Example: A vapor-compression refrigeration cycle uses refrigerant 134a. It enters the compressor at 20psi and leaves at 200psi. The mass flow rate is 0.02lbm/s. Find the power required by the compressor, the cooling rate of the refrigerator, and the COP and Carnot COP.

At state 1, we have saturated vapor at p_1=20psi. From the 134a refrigerant table A-6 in Appendix A, we get h_1=101.39BTU/lbm, s_1=0.2228BTU/lbm, and T_1=-2.405°F.

At state 2, we have p_2=200psi and s_2=s_1=0.2228BTU/lbm. From the 134a refrigerant table A-7, we get h_2=122.08BTU/lbm and T_2=141.5°F.

At state 3, we have saturated liquid at p_3=p_2=200psi. From the 134a refrigerant table A-6, we get h_3=53.76BTU/lbm, and T_3=125.26°F.

At state 4, we have p_4= p_1=20psi, h_4= h_3=53.76BTU/lbm, and T_4= T_1=-2.405°F.

Then

$$\dot{Q}_L = \dot{m}(h_1 - h_4) = 0.02 lbm / s(101.39 BTU / lbm - 53.76 BTU / lbm) = 0.9526 BTU / s$$

$$\dot{Q}_L = 0.9526 BTU / s \left(\frac{3600s}{1hr} \right) \left(\frac{1ton}{12000 BTU / hr} \right) = 0.2858 ton$$

$$\dot{W}_{in} = \dot{m}(h_2 - h_1) = 0.02 lbm/s(122.08 BTU/lbm - 101.39 BTU/lbm) = 0.4138 BTU/s$$

$$\dot{W}_{in} = 0.4138 BTU/s\left(\frac{1hp}{0.7068 BTU/s}\right) = 0.585hp$$

$$COP_R = \frac{\dot{Q}_L}{\dot{W}_{in}} = \frac{0.9526 BTU/s}{0.4138 BTU/s} = 2.30$$

$$COP_{RCarnot} = \frac{1}{T_H/T_L - 1} = \frac{1}{601.5°R/457.6°R - 1} = 3.18$$

11.3.7 Heat pump

The efficiency of a heat pump is also given in terms of a coefficient of performance as

$$COP_{HP} = \frac{1}{1 - Q_L/Q_H}$$

For the idealized case of a Carnot cycle

$$COP_{carnot} = \frac{1}{1 - T_L/T_H}$$

The heat pump process is similar to the refrigeration process except that the outside environment is the cold space, and the warm space is the space needing to be heated. The coefficient of performance can also be expressed as

$$COP_{HP} = \frac{\dot{Q}_H}{\dot{W}_{in}} = \frac{h_2 - h_3}{h_2 - h_1}$$

Example: Heat pumps are used to heat houses in California and New York to maintain them at a temperature of 21°C. If the outside temperature is 10°C in California and 1°C in New York, which heat pump has the highest COP.

For California

$$COP = \frac{1}{1 - 283°K/294°K} = 26.7$$

For New York

$$COP = \frac{1}{1 - 274°K/294°K} = 14.7$$

11.4 Air-Water Vapor Mixture (Humidity)

The atmosphere is a mixture of air and water vapor. Atmospheric pressure is 14.7psi. The pressure is a combination of the partial pressure of the air p_a and partial pressure of the water vapor p_v ; i.e.,

$$p = p_a + p_v$$

The dew point temperature is the temperature at which water vapor begins to condense. The dry bulb temperature T_{db} is the temperature that would be measured by an ordinary thermometer (i.e., the usual measurement of temperature). The wet bulb temperature T_{wb} is the temperature that would be measured by a thermometer encased in a wet material at its saturation point. The specific humidity ω is the mass of water vapor divided by the mass of air; i.e.,

$$\omega = \frac{m_v}{m_a}$$

Then, from the ideal gas law we get $\quad \omega = 0.622 \dfrac{p_v}{p_a}$

The relative humidity φ is the mass of water vapor divided by the mass of water vapor needed to produce a saturated mixture at that temperature; i.e.,

$$\varphi = p_v / p_{satd} = \frac{\omega p_a}{0.622 p_{satd}}$$

where p_{satd} is the pressure of saturated water vapor at the dry bulb temperature.

The partial pressure of the water vapor can be determined from the dry bulb and wet bulb temperatures as

$$p_v = p_{satw} - \frac{(p - p_{satw})(T_{db} - T_{wb})}{2800°F - T_{wb}}$$

where p_{satw} is the pressure of saturated water vapor at the wet bulb temperature.

Example: Water mass calculation
A 10ftx10ftx8ft room is at 80°F, 70% relative humidity, and atmospheric pressure. Find the mass of water vapor in the room.

From steam table A-1 at 80°F, p_{satd}=0.5075psi.

Then

$$\varphi = \frac{p_v}{p_{satd}} \rightarrow p_v = \varphi p_{satd} = 0.7(0.5075 psi) = 0.3553 psi$$

The humidity ratio is

$$\omega = 0.622 \frac{p_v}{p_a} = 0.622 \frac{p_v}{p - p_v} = 0.622 \frac{0.3553 psi}{14.7 psi - 0.3553 psi} = 0.0154$$

The partial pressure of the air is

$$p_a = p - p_v = 14.7 psi - 0.3553 psi = 14.34 psi$$

Since air can be treated as a perfect gas, we can write the mass of air as

$$m_a = \frac{p_a V}{R_a T} = \frac{(14.34 psi)(800 ft^3)}{(0.06855 BTU / lbm°R)(540°R)} \left(\frac{1 BTU}{778 ft lb} \right) \left(\frac{12 in}{1 ft} \right)^2 = 57.36 lbm$$

The mass of water vapor is then

$$m_v = \omega m_a = 0.0154(57.36 lbm) = 0.883 lbm$$

Example: Relative humidity calculation
On a 100°F day, the wet bulb temperature is 70°F. Find the relative humidity.

From steam table A-1 at T_{WB}=70°F, p_{satw}=0.3634psi.
From steam table A-1 at T_{DB}=100°F, p_{satd}=0.9505psi.

Then

$$p_v = p_{satw} - \frac{(p - p_{satw})(T_{db} - T_{wb})}{K - T_{wb}}$$

$$p_v = 0.3634 psi - \frac{(14.7 psi - 0.3634 psi)(100°F - 70°F)}{2800°F - 70°F} = 0.2056 psi$$

The relative humidity is

$$\varphi = \frac{p_v}{p_{satd}} = \frac{0.2056 psi}{0.9505 psi} = 21.6\%$$

11.4.1 Psychrometric charts

When the air-water vapor mixture is at atmospheric pressure (14.7psi), the quantities
described above can be determined from a psychrometric chart. This chart has multiple
scales as shown in Figure 11-9. The dry-bulb temperature T_{db} is represented on the
horizontal axis. The specific humidity ω is represented on the vertical axis. The
wet=bulb temperature T_{wb} is represented by diagonal lines running from the upper left to
lower right. The specific volume v_a is represented by steeper lines running from the upper
left to lower right. Relative humidity φ is represented by curved lines that end at 100%
on the upper left boundary of the chart (This is also the saturation temperature). The
lines of constant humidity ratio (vertical axis) are also lines of constant partial pressure of
water vapor. The lines of constant wet-bulb temperature are also approximately lines of
constant mixture enthalpy $h_{mix}=h_a+\omega h_v$.

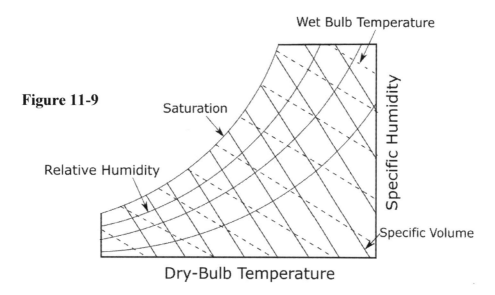

Figure 11-9

A knowledge of two of these quantities is sufficient to determine the others. For
example, consider the last example with $T_{db}=100°F$ and $T_{wb}=70°F$. Using the chart in
Figure 11-10, the intersection of the T_{db} and T_{wb} lines falls between $\varphi=20\%$ and $\varphi=40\%$,
with $\varphi\sim22\%$. Also at this point, $\omega=0.0089$, $h=34BTU/hr$, and $v_a=14.3ft^3/lb$.

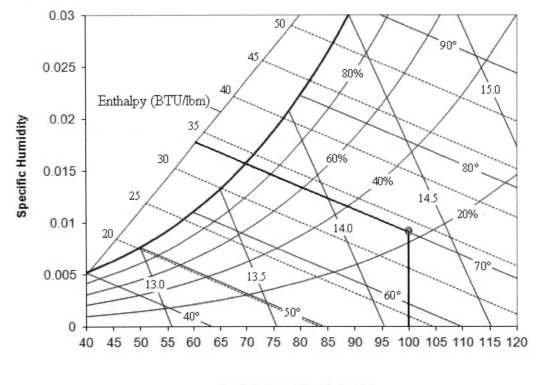

Figure 11-10

11.4.2 Air conditioning

Air conditioning deals with the control of the temperature and humidity of moist air in a confined space. This process involves heating or cooling combined with humidifying or dehumidifying.

11.4.2.1 Heating or cooling without change in moisture content

We consider the case where moist air flows through a heat exchanger without moisture being added or removed as indicated in Figure 11-11.

Figure 11-11
$$\dot{Q}$$
$$\dot{m}_i \qquad \dot{m}_o = \dot{m}_i$$
$$\omega_i \longrightarrow \qquad \longrightarrow \omega_o = \omega_i$$
$$h_i \qquad h_o \neq h_i$$

Since the specific humidity is constant during this process, the change from the incoming condition to the outgoing condition is a horizontal line on psychrometric chart as shown in Figure 11-12.

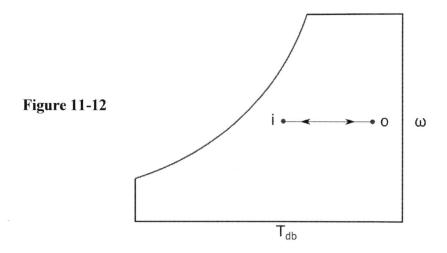

Figure 11-12

The first law gives

$$\dot{Q}_i + \dot{m}_i h_i = \dot{m}_o h_o$$

Example: Heating a room on a cold day
Air at 60°F and 60% relative humidity flows at 1000ft³/min and is heated to 90°F. Find the required heat flow rate and the resulting relative humidity.

First we find $\dot{m}_i$. From the chart in Figure 11-13, v_i~13.2ft³/lbm .

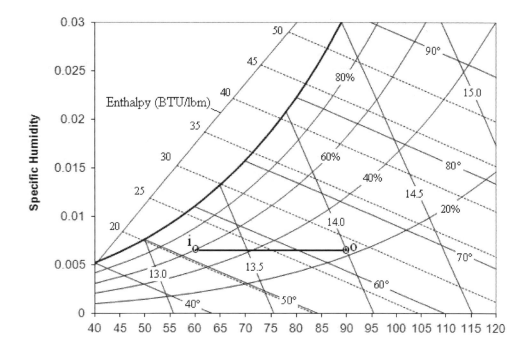

Figure 11-13

Then

$$\dot{m}_i = \dot{V}_i \rho_i = \frac{\dot{V}_i}{v_i} = \frac{1000 \, ft^3/min}{13.2 \, ft^3/lbm}\left(\frac{60 \, min}{1 hr}\right) = 4545 lbm/hr$$

Also from the chart, $h_i \sim 21.5$BTU/lbm, $h_o \sim 29.0$BTU/lbm, and $\varphi_o \sim 22\%$. The energy balance gives

$$\dot{Q}_i = \dot{m}_i(h_o - h_i) = 4545 lbm/hr(29.0 BTU/lbm - 21.5 BTU/lbm) = 34,000 BTU/hr$$

11.4.2.2 Heating or cooling with change in moisture content

We now consider the case where moist air flows through a heat exchanger, and moisture is added or removed as indicated in Figure 11-14.

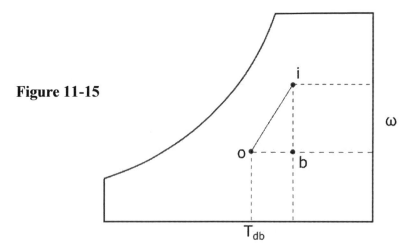

Figure 11-14

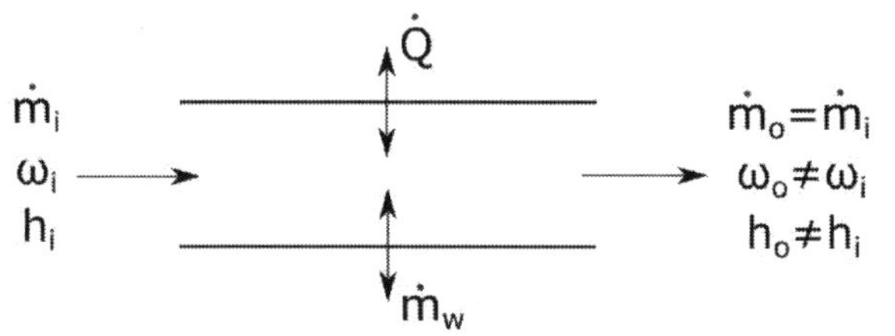

In cooling a room on a hot day, moist air is typically cooled to a temperature below its dew point, and the specific humidity ω becomes lower. The change in condition for this case is illustrated on the psychrometric chart in Figure 11-15.

Figure 11-15

In this case there is a sensible heat transfer that involves the decrease in dry bulb temperature and a latent heat transfer that involves the change in humidity ratio. The energy lost in the condensate $\dot{m}_w$ is usually small enough to be ignored. The sensible heat transfer rate is

$$\dot{Q}_S = \dot{m}_i (h_o - h_b)$$

The latent heat transfer rate is

$$\dot{Q}_L = \dot{m}_i (h_b - h_i)$$

The total heat transfer rate is

$$\dot{Q}_i = \dot{Q}_S + \dot{Q}_L = \dot{m}_i (h_o - h_b) + \dot{m}_i (h_b - h_i) = \dot{m}_i (h_o - h_i)$$

Example: Cooling a room on a hot day
Air at 90°F and 80% relative humidity flows through a chiller at 1000ft³/min and leaves at 70°F and 60% relative humidity. Find the required heat flow rate.

First find $\dot{m}_i$. From the chart in Figure 11-16, $v_i \sim 14.4 \text{ft}^3/\text{lbm}$.

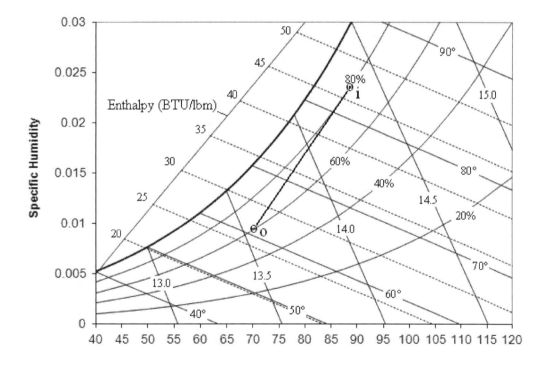

Figure 11-16

Then

$$\dot{m}_i = \dot{V}_i \rho_i = \frac{\dot{V}_i}{v_i} = \frac{1000 \, ft^3 / \min}{14.4 \, ft^3 / lbm} \left(\frac{60 \min}{1 hr} \right) = 4167 lbm / hr$$

Also from the chart, $h_i{\sim}47$BTU/lbm and $h_o{\sim}26$BTU/lbm. The energy balance gives

$$\dot{Q}_i = \dot{m}_i(h_o - h_i) = 4167 lbm / hr(26 BTU / lbm - 47 BTU / lbm) = -87,500 BTU / hr$$

The negative sign indicates that heat has been removed rather than added.

12. FLUID MECHANICS

12.1 Fluid Statics

Consider a stationary fluid as shown in Figure 12-1. The pressure increases with depth according to

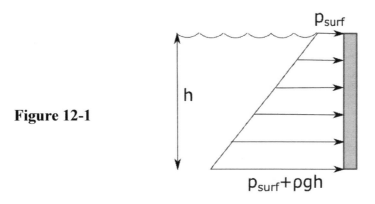

Figure 12-1

$$p = p_{surf} + \rho g h$$

where p_{surf} is the surface pressure (for example, atmospheric pressure), ρ is density, and h is the depth below the surface. The resultant force acting on a tilted plate is

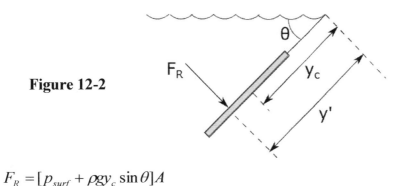

Figure 12-2

$$F_R = [p_{surf} + \rho g y_c \sin \theta] A$$

where y_c is the tilted distance to the centroid of the plate as shown in Figure 12-2, and A is the area of the plate. The force acts at the center of pressure y' given by

$$y' = y_c + \frac{I_c \rho g \sin \theta}{F_R}$$

where I_c is the area moment of inertia about the centroid. A submerged body experiences a buoyancy force equal to the weight of the water displaced by the body; i.e.,

$$F_B = \rho g V$$

where ρ is the density of the fluid, and V is the volume of the submerged body.

Example: A 10ft by 10ft hinged gate on a tank full of water (ρ=62.4lbm/ft³) is held up by force F_1 as shown in Figure 12-3. Find F_1.

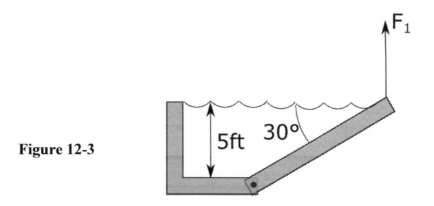

Figure 12-3

Atmospheric pressure acts on the surface of the water and on the back face of the hinged gate. Its effect cancels out and can be ignored. Therefore,

$$y_c = 5\,ft \qquad A = (10\,ft)(10\,ft) = 100\,ft^2 \qquad I_c = \frac{L^4}{12} = \frac{(10\,ft)^4}{12} = 833\,ft^4$$

$$F_R = \rho g y_c \sin\theta A = 62.4\,lb/ft^3(5\,ft)\sin 30°(100\,ft^2) = 15,600\,lb$$

$$y' = y_c + \frac{I_c \rho g \sin\theta}{F_R} = 5\,ft + \frac{(833\,ft^4)(62.4\,lb/ft^3)\sin 30°}{15,600\,lb} = 6.67\,ft$$

We now draw a free body digram of the gate as shown in Figure 12-4.

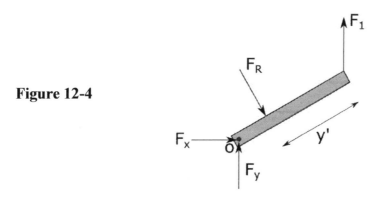

Figure 12-4

Considering moment equilibrium about point o gives

$$\sum M_{pointo} = 0 \rightarrow -F_R(L - y') + F_1 L\cos 30° = 0$$

Solving for F_1 gives

$$F_1 = \frac{F_R(L - y')}{L\cos 30°} = \frac{15,600\,lb(10\,ft - 6.67\,ft)}{10\,ft\cos 30°} = 6,000\,lb$$

12.2 Steady Flow of Incompressible Fluid

Mass balance requires that the total mass flowing into the control volume be equal to the total mass flowing out of the control volume.

$$\sum \dot{m}_i = \sum \dot{m}_o \rightarrow \sum \rho_i \upsilon_i A_i = \sum \rho_o \upsilon_o A_o$$

where υ is flow velocity, and A is cross section area. A very common situation is one where the density is constant (i.e., an incompressible fluid). Then

$$\sum \upsilon_i A_i = \sum \upsilon_o A_o$$

For the case of one inlet and one outlet,

$$\upsilon_i A_i = \upsilon_o A_o$$

12.3 Bernoulli Equation

Recall the first law of thermodynamics for steady flow which states that the total energy into the control volume equals the total energy out; i.e.,

$$\dot{Q}_i + \dot{W}_i + \sum \dot{m}_i (u_i + p_i v_i + \frac{\upsilon_i^2}{2} + gz_i) = \dot{Q}_o + \dot{W}_o + \sum \dot{m}_o (u_o + p_o v_o + \frac{\upsilon_o^2}{2} + gz_o)$$

Consider the case where there are no thermal effects ($\dot{Q}_i = 0$, $\dot{Q}_o = 0$, $u_i = u_o = 0, v_i = 1/\rho_i$, $v_o = 1/\rho_o$). Then

$$\dot{W}_i + \sum \dot{m}_i (p_i / \rho_i + \frac{\upsilon_i^2}{2} + gz_i) = \dot{W}_o + \sum \dot{m}_o (p_o / \rho_o + \frac{\upsilon_o^2}{2} + gz_o)$$

For the case of no work in or out, one inlet and one outlet, and constant density ρ,

$$p_i / \rho + \frac{\upsilon_i^2}{2} + gz_i = p_o / \rho + \frac{\upsilon_o^2}{2} + gz_o$$

or $p/\rho + \dfrac{v^2}{2} + gz = \text{constant}$

This is Bernoulli's equation. We can rewrite this so that each term has units of pressure

$$p + \rho\dfrac{v^2}{2} + \rho gz = \text{constant}$$

We can define stagnation pressure as

$$p_s = p + \rho\dfrac{v^2}{2}$$

where p is the static pressure. If p_s and p are measured by a Pitot tube, then the fluid velocity is determined as

$$v = \sqrt{\dfrac{2(p_s - p)}{\rho}}$$

We can also rewrite the Bernoulli's equation so that each term has units of height as

$$\dfrac{p}{\rho g} + \dfrac{v^2}{2g} + z = \text{constant}$$

The term p/ρg is called the pressure head, $\dfrac{v^2}{2g}$ is the velocity head, and z is the elevation head.

Example: Water flow from a tank
Consider a large tank with a small opening near the bottom as shown in Figure 12-5. Find the velocity of water out of the opening.

Figure 12-5

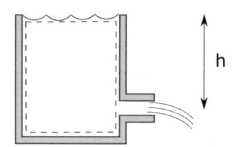

For the control volume, imagine that the top surface of the tank is the inlet, and the opening near the bottom is the outlet. Since A_i is much larger than A_o, then v_i is much smaller than v_o, and it will be ignored in the equation giving

$$\frac{p_i}{\rho g} + z_i = \frac{p_o}{\rho g} + \frac{v_o^2}{2g} + z_o$$

Since $p_i = p_o =$ atmospheric pressure, we get

$$\frac{v_o^2}{2g} = z_i - z_o = h \quad \text{or} \quad v_o = \sqrt{2gh}$$

12.4 Fluid Flow and Work

Let's consider the case where work is input by a pump, work is output by a turbine, and there are friction losses.

$$\dot{m}(p_i/\rho + \frac{v_i^2}{2} + gz_i) + \dot{W}_{pump} = \dot{m}(p_o/\rho + \frac{v_o^2}{2} + gz_o) + \dot{W}_{turb} + \dot{E}_{loss}$$

If we divide by $\dot{m}g$, then every term has units of head h (note that "h" denotes head not enthalpy in fluid mechanics); i.e.,

$$\frac{p_i}{\rho g} + \frac{v_i^2}{2g} + z_i + h_{pump} = \frac{p_o}{\rho g} + \frac{v_o^2}{2g} + z_o + h_{turb} + h_{loss}$$

When pounds mass (lbm) are used for units of mass, it frequently becomes necessary to use the following unit conversion when dealing with force: $1lb = 32.2 lbm\,ft/s^2$.

Example: Losses in a pump
Water enters a pump at 14.7psi and exits at 50psi. A 20kW motor drives the pump. If the flow rate of water is $2ft^3/s$, find the efficiency of the pump.

For our case $z_i = z_o$ and $v_i = v_o$. Let us compute the power output assuming no losses. Then

$$\dot{m}\frac{p_i}{\rho} + \dot{W}_{pump} = \dot{m}\frac{p_o}{\rho}$$

$$\dot{W}_{pump} = \frac{\dot{m}}{\rho}(p_o - p_i) = \dot{V}(p_o - p_i)$$

$$\dot{W}_{pump} = (2\,ft^3/s)(50lb/in^2 - 14.7lb/in^2)\left(\frac{12in}{1\,ft}\right)^2 = 10,166\,ftlb/s\left(\frac{1kW}{738\,ftlb/s}\right) = 13.78kW$$

Since the motor delivers 20kW to the pump, while the pump output is only 13.78kW, the difference between these is the result of loss in the pump. Therefore, the efficiency is

$$\eta = \frac{13.78kW}{20kW} = 69\%$$

Example: Hydroelectric power
Water with $\rho = 62.4lbm/ft^3$ drops from an elevation of 300ft and flows through a turbine at $1000ft^3/s$. If the electrical output is 20MW, find the efficiency of this power system.

We will assume that the energy is solely the result of the change in elevation; i.e., the change in flow velocity is negligible. Let's compute the power output of the turbine if there are no losses.

$$\dot{m}gz_i = \dot{m}gz_o + \dot{W}_{turb}$$
$$\dot{W}_{turb} = \dot{m}gz_i - \dot{m}gz_o = \rho\dot{V}g(z_i - z_o)$$
$$\dot{W}_{turb} = (62.4lbm/ft^3)(1000\,ft^3/s)(32.2\,ft/s^2)(300\,ft)\left(\frac{1lb}{32.2lbmft/s^2}\right) = 18.7x10^6\,ftlb/s$$

$$\dot{W}_{turb} = 18.7x10^6\,ftlb/s\left(\frac{1kW}{738\,ftlb/s}\right) = 25.37MW$$

The electrical output is less than the mechanical energy available. Therefore, the efficiency is

$$\eta = \frac{\dot{W}_{electic}}{\dot{W}_{turb}} = \frac{20MW}{25.37MW} = 78.8\%$$

12.5 Fluid Forces

Mass balance requires that the total mass flowing into the control volume be equal to the total mass flowing out of the control volume.

$$\sum \dot{m}_i = \sum \dot{m}_o \rightarrow \sum \rho v_i A_i = \sum \rho v_o A_0$$

Newton's second law applied to the control volume gives

$$\sum F_x = \sum \dot{m}_o v_{ox} - \sum \dot{m}_i v_{ix}$$

$$\sum F_y = \sum \dot{m}_o \upsilon_{oy} - \sum \dot{m}_i \upsilon_{iy}$$

where F_x and F_y are the force components in the x and y directions acting on the control volume.

Example: Spray from a hose
Water from a hose is sprayed against a wall as shown in Figure 12-6. The water velocity is 25ft/s through a nozzle opening with a cross-section area of $0.75in^2$. Find the force on the wall.

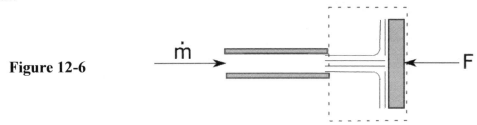

Figure 12-6

The mass flow rate is

$$\dot{m} = \rho \upsilon A = (62.4 lbm / ft^3)(25 ft / s)(0.75 in^2)\left(\frac{1 ft}{12 in}\right)^2 = 8.125 lbm / s$$

Using the control volume shown, we get $\upsilon_{ix} = 25 ft / s$ and $\upsilon_{ox} = 0$.
Therefore,

$$\sum F_x = \dot{m}(\upsilon_{ox} - \upsilon_{ix})$$

$$-F = -(8.125 lbm / s)(25 ft / s)\left(\frac{1 lb}{32.2 lbm ft / s^2}\right)$$

$$F = 6.31 lb$$

Example: Force on a nozzle
Water flows through a hose with a nozzle on the end as shown in Figure 12-7. The water exits the nozzle at a velocity of 40ft/s. Find the force required to hold the nozzle onto the end of the hose.

$D_o = 0.5in$

Figure 12-7 $D_i = 1in$ 40ft/s

First, we calculate the mass flow rate.

$$\dot{m} = \rho A_o \upsilon_o = 62.4 lbm / ft^3 \left[\frac{\pi(0.5 in)^2}{4}\right](40 ft / s)\left(\frac{1 ft}{12 in}\right)^2\left(\frac{1 slug}{32.2 lbm}\right) = 0.106 slug / s$$

To calculate the velocity υ_i, we use the continuity of mass equation.

$$\rho A_i \upsilon_i = \rho A_o \upsilon_o$$

Solving for υ_i gives

$$\upsilon_i = \frac{A_o}{A_i}\upsilon_o = \frac{\pi[(0.5in)^2/4]}{\pi[(1in)^2/4]}(40\,ft/s) = 10\,ft/s$$

The Bernoulli equation gives

$$p_i + \rho\upsilon_i^2/2 = p_{atm} + \rho\upsilon_o^2/2$$

Solving for p_i gives

$$p_i = p_{atm} + \rho(\upsilon_o^2 - \upsilon_i^2)/2$$

$$p_i = 14.7\,psi + 62.4\,lbm/ft^3[(40\,ft/s)^2 - (10\,ft/s)^2)]/2\left(\frac{1\,ft}{12in}\right)^2\left(\frac{1\,slug}{32.2\,lbm}\right) = 24.8\,psi$$

A control volume around the nozzle is shown in Figure 12-8.

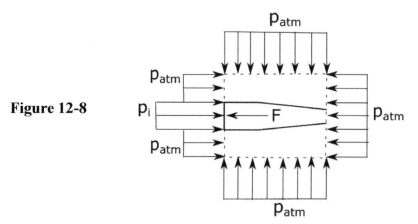

Figure 12-8

Some of the resultant forces from the atmospheric pressure cancel one another out. If we re-draw the control volume with the atmospheric pressure subtracted from all surfaces as shown in Figure 12-9, we get the following.

Figure 12-9 $p_i - p_{atm}$

The force on the nozzle can be determined from

$$\sum F_x = \dot{m}(\upsilon_{ox} - \upsilon_{ix})$$
$$(p_i - p_{atm})A_i - F = \dot{m}(\upsilon_{ox} - \upsilon_{ix})$$

Solving for F gives

$$F = (24.8psi - 14.7psi)[\pi(1in)^2 / 4] - 0.106 slug / s(40 ft / s - 10 ft / s) = 4.75 lb$$

12.6 Flow in Conduits

Bernoulli's equation from the previous section ignores viscous effects in the fluid. These effects are similar to friction effects that arise when one body slides over another. When a fluid flows past a stationary solid surface, the fluid tends to stick to the surface. This results in a non-uniform velocity in the fluid. The fluid velocity profile looks like that shown in Figure 12-10.

Figure 12-10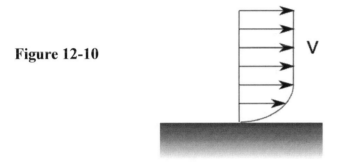

The viscous effects produce a shear stress in the fluid as

$$\tau = \mu \frac{d\upsilon}{dy}$$

where μ is the viscosity of the fluid (a material property). These effects are important for fluid flowing through a pipe. We will categorize the type of flow as either laminar (smooth flow) or turbulent (disordered flow). These two types can be identified by the Reynolds number Re.

$$Re = \frac{InertiaForces}{ViscousForces} = \frac{\rho \upsilon_m D}{\mu} = \frac{\upsilon_m D}{\nu}$$

where ρ is density, υ_m is mean velocity (equal to the volumetric flow rate $\dot{V}$ divided by the area; i.e., $\upsilon_m = \dot{V} / (\pi D^2 / 4)$, D is diameter of the pipe, and $\nu = \mu/\rho$ is the kinematic viscosity. The type of flow is determined by the Reynolds number as follows:

If Re< 2100 →laminar flow

If Re>10,000 →turbulent flow

12.6.1 Laminar flow in a circular pipe

The fluid velocity profile in a pipe experiencing laminar flow looks like that shown in Figure 12-11.

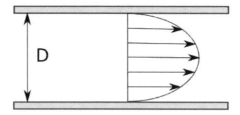

Figure 12-11

The viscous effects cause a pressure drop Δp in the pipe over a length L given by

$$\Delta p = \frac{128 \mu L \dot{V}}{\pi D^4}$$

where again $\dot{V}$ is the volumetric flow rate. The average velocity across the pipe diameter is

$$\upsilon_m = \frac{\dot{V}}{\pi D^2 / 4}$$

Example: Find the pressure drop in oil (with ρ=900kg/m^3 and μ=0.18kg/ms) flowing at 0.01m^3/s through 10m of pipe with a diameter of 0.1m.

First, we calculate the Reynolds number

$$\mathrm{Re} = \frac{\rho \upsilon_m D}{\mu} = \frac{\rho \dot{V} D}{\left(\frac{\pi D^2}{4}\right) \mu} = \frac{(900 kg/m^3)(0.01m^3/s)(0.1m)}{\left(\frac{\pi (0.1m)^2}{4}\right)(0.18 kg/ms)} = 637$$

Since Re<2100, we confirm that the flow is laminar. The pressure drop is

$$\Delta p = \frac{128 \mu L \dot{V}}{\pi D^4} = \frac{128(0.18 kg/ms)(10m)(0.01m^3/s)}{\pi (0.1m)^4} = 7,334 Pa$$

12.6.2 General flow in a pipe

For the case of laminar or turbulent flow in a circular or non-circular pipe, the pressure loss with distance is

$$\Delta p = f \frac{L}{D} \rho \frac{\upsilon_m^2}{2}$$

where f is the friction factor, and D is diameter if the pipe is circular. If the pipe has a rectangular cross-section with dimensions aXb, then D is replaced by 2ab(a+b).

If we divide the equation above by ρg to convert it into units of height (head), then the head loss is

$$h_{loss} = \frac{\Delta p}{\rho g} = f \frac{\upsilon_m^2}{2g} \frac{L}{D}$$

The friction factor f is a function of the Reynolds number Re and the relative roughness ε/D. The value of f can be read off a Moody Diagram shown in Figure 12-12.

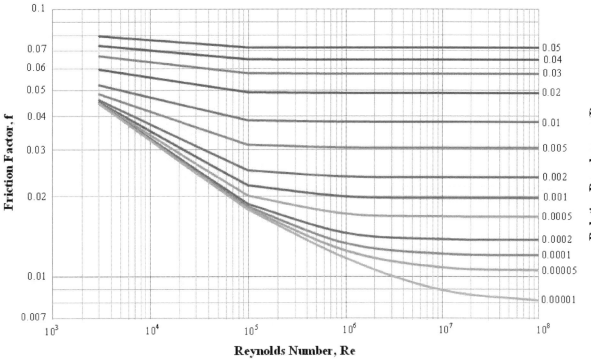

Figure 12-12 Moody Diagram

The valves, elbows, etc. in a pipe system cause additional pressure losses. The so-called "minor losses" are often expressed in terms of an equivalent length of pipe L_{eq} , so that

$$h_{loss-minor} = f \frac{\upsilon_m^2}{2g} \frac{L_{eq}}{D}$$

The values $h_{loss-minor}$ can be determined for some common cases through

$$h_{loss-minor} = K \frac{\upsilon^2}{2g}$$

where the value of K is given in Figure 12-13.

Figure 12-13

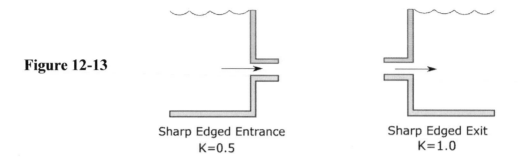

Sharp Edged Entrance
K=0.5

Sharp Edged Exit
K=1.0

The total head loss is the summation of all these effects.

$$h_{loss-total} = \sum h_{loss-major} + \sum h_{loss-minor}$$

12.6.3 Pipe networks and pumps

Pipes can be connected in series or parallel. For pipes connected in series, the flow rate is uniform throughout the system. The total pressure loss (or head loss) is equal to the sum of the losses in each section. For pipes connected in parallel, the total flow rate is equal to the sum of the flow rates in each branch. The pressure loss (or head loss) in each branch must be the same because the pressure in each branch at the junctions must be the same. With pumps and turbines in the network, we have

$$p_i / \rho + \frac{v_i^2}{2} + g z_i + \frac{\dot{W}_{pump}}{\dot{m}} = p_o / \rho + \frac{v_o^2}{2} + g z_o + \frac{\dot{W}_{turbine}}{\dot{m}} + g h_{loss}$$

or $\qquad \dfrac{p_i}{\rho g} + \dfrac{v_i^2}{2g} + z_i + h_{pump} = \dfrac{p_o}{\rho g} + \dfrac{v_o^2}{2g} + z_o + h_{turbine} + h_{loss}$

Example: Water enters the system at a rate of 10gal/min at position 1 and exits at position 2 as shown in Figure 12-14. The two branches are identical except that the equivalent length of branch A is 100ft and the equivalent length of branch B is 150ft. Find the flow rate in each branch.

Branch A

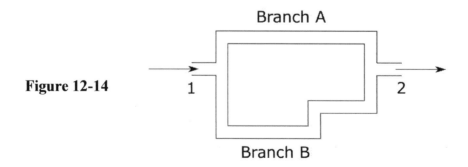

Figure 12-14   1   2

Branch B

$$h_{lossA} = f \frac{\upsilon_A^2}{2g} \frac{L_{Aeq}}{D}$$

$$h_{lossB} = f \frac{\upsilon_B^2}{2g} \frac{L_{Beq}}{D}$$

For a parallel system $\quad h_{lossA} = h_{lossB} \rightarrow f \frac{\upsilon_A^2}{2g} \frac{L_{Aeq}}{D} = f \frac{\upsilon_B^2}{2g} \frac{L_{Beq}}{D}$

$$\upsilon_B = \sqrt{\frac{L_{Aeq}}{L_{Beq}}} \upsilon_A$$

Since $\quad \upsilon = \dfrac{\dot{V}}{A}$, then $\quad \dot{V}_B = \sqrt{\dfrac{L_{Aeq}}{L_{Beq}}} \dot{V}_A$

Mass balance $\quad \dot{m}_{total} = \dot{m}_A + \dot{m}_B \rightarrow \dot{V}_{total} = \dot{V}_A + \dot{V}_B$

$$\dot{V}_{total} = \dot{V}_A + \sqrt{\frac{L_{Aeq}}{L_{Beq}}} \dot{V}_A$$

$$10 = \dot{V}_A + \sqrt{\frac{100}{150}} \dot{V}_A \rightarrow \dot{V}_A = 5.5 \, gal/\min$$

$$\dot{V}_B = \dot{V}_{total} - \dot{V}_A = 10 \, gal/\min - 5.5 \, gal/\min = 4.5 \, gal/\min$$

Example: Pump requirements to overcome flow losses
An industrial process requires 1ft³/s of water at 60°F (with μ=2.713lbm/fthr) to be pumped through a 10ft long pipe with a diameter of 3in (0.25ft) as shown in Figure 12-15. How much of the pump's power is required simply to overcome losses?

Figure 12-15

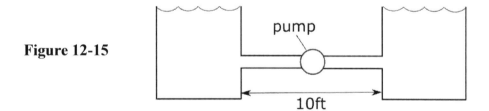

The losses arise from three sources: a sharp edged pipe entrance, a sharp edged pipe exit, and the roughness of 10 feet of pipe. The flow velocity is

$$\upsilon = \dot{V}/A = (1 ft^3/s)/[\pi(0.25 ft)^2/4] = 20.4 \, ft/s$$

The Reynolds number is

$$\text{Re} = \frac{\rho \upsilon^2 D}{\mu} = \frac{(62.36 lbm / ft^3)(20.4 ft / s)^2 (0.25 ft)}{(2.713 lbm / ft hr)\left(\dfrac{1hr}{3600s}\right)} = 8.63x10^6$$

For commercial steel $\varepsilon = 0.00015 ft$
Then, $\varepsilon/D = 0.00015 ft / 0.25 ft = 0.0006$
From the moody chart, $f = 0.0175$

The head loss from 10 feet of pipe is

$$h_{loss-pipe} = f\frac{L\upsilon_m^2}{2gD} = 0.0175\frac{(10 ft)(20.4 ft / s)^2}{2(32.2 ft / s^2)(0.25 ft)} = 4.52 ft$$

The head loss from the entrance (with K=0.5) is

$$h_{loss-entrance} = K\frac{\upsilon_m^2}{2g} = 0.5\frac{(20.4 ft / s^2)^2}{2(32.2 ft / s^2)} = 3.23 ft$$

The head loss from the exit (with K=1.0) is

$$h_{loss-exit} = K\frac{\upsilon_m^2}{2g} = 1.0\frac{(20.4 ft / s)^2}{2(32.2 ft / s^2)} = 6.46 ft$$

The total head loss is

$$h_{loss} = h_{loss-pipe} + h_{loss-entrance} + h_{loss-exit} = 14.21 ft$$

The power required simply to overcome these losses alone is

$$\frac{\dot{W}_{pump}}{\dot{m}} = gh_{loss}$$

$$\dot{W}_{pump} = \dot{m}gh_{loss} = \rho \dot{V}gh_{loss} = (62.36 lbm / ft^3)(1 ft^3 / s)(32.2 ft / s^2)(14.21 ft)\left(\frac{1lb}{32.2 lbmft / s^2}\right)$$

$$\dot{W}_{pump} = 886 ft lb / s\left(\frac{1hp}{550 ft lb / s}\right) = 1.61 hp$$

12.7 Flow Around Bodies (Lift and Drag)

The combination of pressure forces and viscous forces produces a resultant force on an object when fluid flows around it as shown in Figure 12-16.

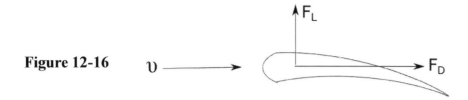

Figure 12-16

The component of force that is parallel to the flow direction is called the drag force F_D which can be expressed as

$$F_D = C_D(\frac{1}{2}\rho v^2 A_P)$$

where C_D is the drag coefficient, and A_P is the projected area of the body. The component of force that is perpendicular to the direction of flow is called the lift force F_L which can be expressed as

$$F_L = C_L(\frac{1}{2}\rho v^2 A_P)$$

where C_L is the lift coefficient.

For the general case of a smooth cylinder, the drag coefficient is given in Figure 12-17.

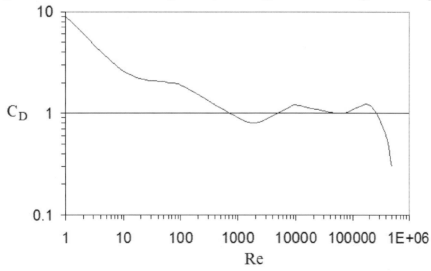

Figure 12-17

Example: Wind force on a pole
A 6in diameter pole that is 30ft high is subjected to a wind at 30mph at 80°F with ρ=0.074lbm/ft^3 and v=0.17x10^{-3}ft^2/s. Find the drag force from the wind.

The wind velocity is

$$\upsilon = 30 mile/hr\left(\frac{5280\,ft}{1mile}\right)\left(\frac{1hr}{3600s}\right) = 44\,ft/s$$

The Reynolds number for the air is

$$Re = \frac{\upsilon D}{\nu} = \frac{(44\,ft/s)(0.5\,ft)}{0.17x10^{-3}\,ft^2/s} = 1.29x10^5$$

From the figure for a smooth cylinder, we have C_D=1.1. The drag force is

$$F_D = C_D(\frac{1}{2}\rho\upsilon^2 A_P) = 1.1\{\frac{1}{2}(0.074lbm/ft^3)(44\,ft/s)^2[(0.5\,ft)(30\,ft)]\}\left(\frac{1lb}{32.2lbmft/s^2}\right)$$

$$F_D = 36.7lb$$

12.8 Compressible Flow

Up to this point, we have treated the fluid as having constant density (i.e., incompressible). For high-speed flow of a gas, compressibility effects become important, and we will use the ideal gas law.

$$pv = RT \qquad or \qquad p = \rho RT$$

In a compressible fluid, disturbances in pressure propagate through the fluid as sound waves with velocity given as

$$c = \sqrt{kRT}$$

with k=C_p/C_V. We can express the flow velocity of the fluid υ in terms of the Mach number given by

$$M = \upsilon/c$$

with

M<1 $\rightarrow$ subsonic flow
M>1 $\rightarrow$ supersonic flow

The temperature, pressure, and density of the flowing fluid can be expressed in terms of their usual static values T, p, ρ or in terms of their stagnation values T_s, p_s, ρ_s which are the values they would take on if the velocity were brought to zero. These quantities are related by

$$\frac{T_s}{T} = 1 + \frac{k-1}{2}M^2$$

$$\frac{p_s}{p} = \left(1 + \frac{k-1}{2} M^2\right)^{k/(k-1)}$$

$$\frac{\rho_s}{\rho} = \left(1 + \frac{k-1}{2} M^2\right)^{1/(k-1)}$$

Example: Supersonic airplane
An airplane flies at M=2 in air at a temperature of T=-30°F. Find the skin temperature of the airplane.

The skin temperature should be close to the stagnation temperature. For air at -30°F, k~1.4.

$$T_o = T\left(1 + \frac{k-1}{2} M^2\right) = 430°R\left(1 + \frac{1.4-1}{2}(2)^2\right) = 774°R = 314°F$$

12.8.1 Normal shock waves

Shock waves occur when a fluid is flowing at a supersonic velocity or if an object is moving at supersonic velocity through a stationary fluid. A normal shock wave is a wave front across which fluid is decelerated from supersonic velocity M_1 to subsonic velocity M_2. There will be a jump in flow velocity as well as static temperature, pressure, and density across the wave front given as

$$M_2 = \sqrt{\frac{(k-1)M_1^2 + 2}{2kM_1^2 - (k-1)}}$$

$$\frac{T_2}{T_1} = \frac{1 + [(k-1)/2]M_1^2}{1 + [(k-1)/2]M_2^2}$$

$$\frac{p_2}{p_1} = \frac{1 + kM_1^2}{1 + kM_2^2}$$

$$\frac{\rho_2}{\rho_1} = \left(\frac{1 + kM_1^2}{1 + kM_2^2}\right)\left(\frac{1 + [(k-1)/2]M_1^2}{1 + [(k-1)/2]M_2^2}\right)$$

The stagnation temperature is unchanged across the shockwave. However, the stagnation pressure change is given by

$$\frac{p_{s2}}{p_{s1}} = \frac{\left(\dfrac{[(k+1)/2]M_1^2}{1 + [(k-1)/2]M_1^2}\right)^{k/(k-1)}}{\left([2k/(k+1)]M_1^2 - (k-1)/(k+1)\right)^{1/(k-1)}}$$

Example: Normal shock wave

The Mach number upstream of a normal shock wave in air is 3. The upstream static pressure and temperature are 14.7psi and 70 °F, respectively. Find the static temperature and pressure downstream of the shock wave.

First we calculate the downstream Mach number.

$$M_2 = \sqrt{\frac{(k-1)M_1^2 + 2}{2kM_1^2 - (k-1)}} = \sqrt{\frac{(1.4-1)(3)^2 + 2}{2(1.4)(3)^3 - (1.4-1)}} = 0.475$$

Then for M_2=0.475, we get

$$T_2 = T_1 \left(\frac{1+[(k-1)/2]M_1^2}{1+[(k-1)/2]M_2^2} \right) = (530°R) \left(\frac{1+[(1.4-1)/2](3)^2}{1+[(1.4-1)/2](0.475)^2} \right) = 1420°R = 960°F$$

$$p_2 = p_1 \left(\frac{1+kM_1^2}{1+kM_2^2} \right) = (14.7\,psi) \left(\frac{1+(1.4)(3)^2}{1+(1.4)(0.475)^2} \right) = 152\,psi$$

13. HEAT TRANSFER

13.1 Mechanisms of Heat Transfer

The primary mechanisms of heat transfer are conduction, convection, and radiation. Conduction typically involves heat flow through a solid or a stationary fluid. The rate of heat conduction through a wall of area A and thickness L as shown in Figure 13-1 is

Figure 13-1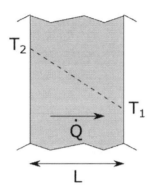

$$\dot{Q}_{cond} = KA\frac{(T_2 - T_1)}{L}$$

where K is the thermal conductivity (a material property - see Table 13-1). Note that heat always flows from the higher temperature region to the lower temperature region.

Table 13-1

Material	Thermal conductivity (W/m K)	Thermal conductivity (BTU/hrftF)
Silver	406	234.5868
Copper	385	222.453
Gold	314	181.4292
Brass	109	62.9802
Aluminum	205	118.449
Iron	79.5	45.9351
Steel	50.2	29.00556
Glass	0.8	0.46224
Concrete	0.8	0.46224
Asbestos	0.08	0.046224
Fiberglass	0.04	0.023112
Brick	0.6	0.34668
Polystyrene (styrofoam)	0.033	0.0190674
Wood	0.1	0.05778

Convection involves heat transfer between a solid surface and a <u>moving</u> fluid. For this case the rate of heat flow is

$$\dot{Q}_{conv} = hA(T_S - T_F)$$

where h is the convective heat transfer coefficient or film coefficient, T_s is the surface temperature of the solid, and T_F is the fluid temperature away from the surface.

Radiation involves heat transfer from a surface by electromagnetic waves. The rate of heat radiated from a body is

$$\dot{Q}_{rad} = \varepsilon\sigma A T_{S1}^4$$

where ε is the emissivity of the surface (see Table 13-2), $\sigma = 5.67 \times 10^{-8} W/m^{2o}K^4 = 0.1714 \times 10^{-8}$ BTU/hrft^{2o}R^4 is the Stefan-Boltzmann constant, and T_{s1} is the absolute temperature of the surface of the body. For a "black body", $\varepsilon=1$.

Table 13-2

Material	Emissiviy
Water	0.95
Aluminum	0.1
Plastic	0.93
Ceramic	0.94
Glass	0.87
Rubber	0.9
Cloth	0.95
Black Paint	0.98
White Paint	0.9

13.2 Conduction

Heat flow is analogous to the flow of electrical current. If we write the heat conduction equation as

$$\dot{Q} = \frac{(T_2 - T_1)}{R} \quad \text{and} \quad R = \frac{L}{KA} = \text{thermal resistance}$$

it is similar to Ohm's law for electric current

$$I = \frac{(V_2 - V_1)}{R} \quad \text{where is R electrical resistance}$$

We can use the laws of resistors in series or parallel for heat transfer. For example, consider a wall between two fluids as shown in Figure 13-2.

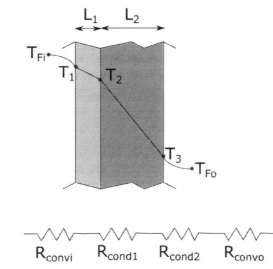

Figure 13-2

We can write $\quad \dot{Q} = \dfrac{(T_{Fi} - T_{Fo})}{R_{total}}$

where $\quad R_{total} = R_{convi} + R_{cond1} + R_{cond2} + R_{convo} = \dfrac{1}{h_i A} + \dfrac{L_1}{K_1 A} + \dfrac{L_2}{K_2 A} + \dfrac{1}{h_o A}$

<u>Example</u>: Heat flow through a uniform wall

Heat flows through a 0.5in thick wooden wall in a house. The inside temperature is 70°F. The outside temperature is 40°F. The thermal conductivity of the wood is 0.1BTU/hrft°F. The convection heat transfer coefficient for the inside is 3BTU/hrft²°F. The convection heat transfer coefficient for the outside is 6BTU/hrft²°F. The wall is 30 feet long and 8 feet high. Find the heat lost through the wall.

First, calculate the resistances

$$R_{convi} = \frac{1}{h_i A} = \frac{1}{3BTU/hrft^2°F(30ft)(8ft)} = 1.39x10^{-3}(BTU/hr°F)^{-1}$$

$$R_{cond} = \frac{L}{KA} = \frac{0.5in(1ft/12in)}{0.1BTU/hrft°F(30ft)(8ft)} = 1.74x10^{-3}(BTU/hr°F)^{-1}$$

$$R_{convo} = \frac{1}{h_o A} = \frac{1}{6BTU/hrft^2°F(30ft)(8ft)} = 0.69x10^{-3}(BTU/hr°F)^{-1}$$

The total resistance is

$$R_{total} = R_{convi} + R_{cond} + R_{convo} = 3.82x10^{-3}(BTU/hr°F)^{-1}$$

The heat flow rate is

$$\dot{Q} = \frac{(T_i - T_o)}{R_{total}} = \frac{(70°F - 40°F)}{3.82x10^{-3}(BTU/hr°F)^{-1}} = 7853 BTU/hr$$

<u>Example</u>: Heat flow through a composite wall
Heat flows through a wall made of alternating sections of wood and insulation material as shown in Figure 13-3. The thermal conductivity of the wood is 0.1BTU/hrft°F. The thermal conductivity of the insulation material is 0.02BTU/hrft°F. The convection heat transfer coefficient for the inside is 3BTU/hrft²°F. The convection heat transfer coefficient for the outside is 6BTU/hrft²°F. The inside temperature is 70°F. The outside temperature is 40°F. Find the rate of heat flow through the section shown by the dashed lines if the height is 8 feet.

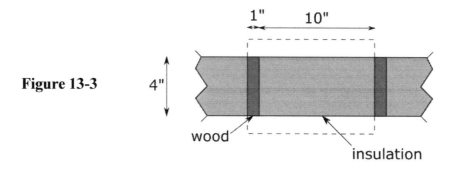

Figure 13-3

First, calculate the resistances

$$R_{convi} = \frac{1}{h_i A} = \frac{1}{3BTU/hrft^2°F(11/12ft)(8ft)} = 0.0455(BTU/hr°F)^{-1}$$

$$R_{convo} = \frac{1}{h_o A} = \frac{1}{6BTU/hrft^2°F(11/12ft)(8ft)} = 0.0227(BTU/hr°F)^{-1}$$

$$R_{condw} = \frac{L_w}{K_w A_w} = \frac{4/12ft}{0.1BTU/hrft°F(1/12ft)(8ft)} = 5.0(BTU/hr°F)^{-1}$$

$$R_{condl} = \frac{L_l}{K_l A_l} = \frac{4/12ft}{0.02BTU/hrft°F(10/12ft)(8ft)} = 2.5(BTU/hr°F)^{-1}$$

Because wood and insulation are in parallel, their combined resistance is

$$\frac{1}{R_{comb}} = \frac{1}{R_{condw}} + \frac{1}{R_{condl}} = \frac{1}{5.0(BTU/hr°F)^{-1}} + \frac{1}{2.5(BTU/hr°F)^{-1}} \rightarrow R_{comb} = 1.67(BTU/hr°F)^{-1}$$

The total heat flow resistance is

$$R_{total} = R_{convi} + R_{comb} + R_{convo} = 1.74(BTU/hr°F)^{-1}$$

The heat flow rate is

$$\dot{Q} = \frac{(T_i - T_o)}{R_{total}} = \frac{(70°F - 40°F)}{1.74(BTU/hr°F)^{-1}} = 17.2 BTU/hr$$

13.2.1 Cylinders and spheres

Heat flow rate through a hollow cylinder is

$$\dot{Q}_{cyl} = \frac{T_i - T_o}{R} \quad \text{where} \quad R = \frac{\ln(D_o/D_i)}{2\pi LK}$$

where D_o is the outer diameter, D_i is the inner diameter, and L is the length of the cylinder.

Heat flow rate through a hollow sphere is

$$\dot{Q}_{sph} = \frac{T_i - T_o}{R} \quad \text{where} \quad R = \frac{D_o - D_i}{2\pi D_i D_o K}$$

Example: Heat loss in a water pipe
Water at 140°F flows in a plastic pipe with an inside diameter of 1in and a wall thickness of 0.125in. The thermal conductivity of the plastic is 0.06BTU/hrft°F. The convection heat transfer coefficient inside the pipe is 8BTU/hrft^2°F. The convection heat transfer coefficient outside the pipe is 4BTU/hrft^2°F. Find the heat loss over 10 feet of pipe if the outside temperature is 70°F.

First, calculate the inside and outside surface areas of the pipe.

$$A_i = \pi D_i L = \pi(1/12\,ft)(10\,ft) = 2.62\,ft^2$$
$$A_o = \pi D_o L = \pi(1.25/12\,ft)(10\,ft) = 3.27\,ft^2$$

Now, calculate the resistances

$$R_{pipe} = \frac{\ln(D_o/D_i)}{2\pi KL} = \frac{\ln(1.25in/1in)}{2\pi(0.06BTU/hrft°F)(10\,ft)} = 0.0592(BTU/hr°F)^{-1}$$

$$R_{convi} = \frac{1}{h_i A_i} = \frac{1}{8BTU/hrft^2°F(2.62\,ft^2)} = 0.0477(BTU/hr°F)^{-1}$$

$$R_{convo} = \frac{1}{h_o A_o} = \frac{1}{4BTU/hrft^2°F(3.27\,ft^2)} = 0.0764(BTU/hr°F)^{-1}$$

The total heat flow resistance is

$$R_{total} = R_{convi} + R_{pipe} + R_{convo} = 0.183(BTU / hr°F)^{-1}$$

The heat flow rate is

$$\dot{Q} = \frac{(T_i - T_o)}{R_{total}} = \frac{(140°F - 70°F)}{0.183(BTU / hr°F)^{-1}} = 382 BTU / hr$$

13.2.2 R-value of insulation

The R-value of insulation is the thermal resistance per unit area. For flat insulation this value is

$$Rvalue = L / K$$

For pipe insulation this value is

$$Rvalue = \frac{D_o}{2K} \ln \frac{D_o}{D_i}$$

Given the R-value, the heat flow rate becomes

$$\dot{Q} = \frac{(T_i - T_o)A}{Rvalue}$$

13.3 Forced Convection

13.3.1 Internal and external flow

The convective heat transfer coefficient is frequently determined by empirical relations between dimensionless numbers. The Nusselt number Nu is defined as

$$Nu = \frac{hL}{K}$$

where L is the characteristic length. The Prandtl number Pr is defined as

$$Pr = \frac{\mu C_P}{K}$$

where μ is the viscosity of the fluid, and C_p is the specific heat. The Reynolds number Re is defined as

$$\mathrm{Re}_x = \frac{\rho \upsilon x}{\mu}$$

where υ is the flow velocity. The relation for turbulent flow ($\mathrm{Re} > 10^4$) in a pipe is

$$Nu = 0.023 \mathrm{Re}_D^{0.8} \mathrm{Pr}^{\frac{1}{3}}$$

The relation for turbulent flow ($\mathrm{Re} > 10^5$) over a flat plate is

$$Nu = 0.036 \mathrm{Re}_L^{0.8} \mathrm{Pr}^{\frac{1}{3}}$$

Example: Water flow over a hot plate
Water at 80°F flows at 10ft/s over a plate at 200°F with a length of 5ft and a width of 2ft. Find the heat transfer rate.

The film temperature is

$$T_f = (T_s + T_\infty)/2 = (200°F + 80°F)/2 = 140°F$$

At 140°F, water has the following properties: $\rho = 61.38 \text{lbm/ft}^3$, $K = 0.378 \text{BTU/hrft°F}$, $\mu = 1.129 \text{lbm/fthr}$, $\mathrm{Pr} = 2.98$. The Reynolds number is

$$\mathrm{Re}_L = \frac{\rho \upsilon L}{\mu} = \frac{(61.38 lbm/ft^3)(10 ft/s)(5 ft)}{1.129 lbm/fthr}\left(\frac{3600s}{1hr}\right) = 9.79x10^6$$

This Reynolds number corresponds to turbulent flow. Therefore,

$$Nu = 0.036 \mathrm{Re}^{0.8} \mathrm{Pr}^{1/3} = 0.036(9.79x10^6)^{0.8}(2.98)^{1/3} = 20{,}270$$

The heat transfer coefficient h is

$$h = \frac{K}{L} Nu = \frac{0.387 BTU/hrft°F}{5 ft}(20{,}270) = 1568 BTU/hrft^2°F$$

The heat transfer rate is

$$\dot{Q} = hA(T_s - T_\infty) = (1568 BTU/hrft^2°F)(5 ft)(2 ft)(200°F - 80°F) = 1.88x10^6 BTU/hr$$

13.3.2 Flow in heat exchangers

Consider a heat exchanger where a hot fluid flows through a pipe while a cold fluid flows through another pipe concentric with the hot fluid pipe as shown in Figure 13-4.

Figure 13-4

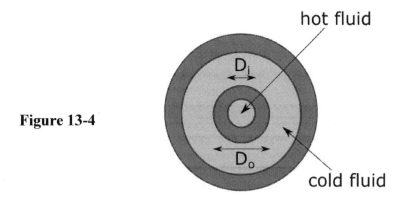

Heat is transferred from the hot fluid to the inside of the pipe by convection, then through the pipe wall by conduction, and finally from the outside pipe wall to the cold fluid by convection. The total thermal resistance is

$$R_{total} = R_{convi} + R_{cond} + R_{convo} = \frac{1}{h_i A_i} + \frac{\ln(D_o / D_i)}{2\pi L K} + \frac{1}{h_o A_o}$$

where

A_i=area of the inner surface of the inside pipe
A_o=area of the outer surface of the inside pipe
D_i=inner diameter of inside pipe
D_o=outer diameter of inside pipe

The heat flow rate is

$$\dot{Q} = \frac{\Delta T}{R_{total}}$$

We can define overall heat transfer coefficients U_i and U_o as

$$U_i = \frac{\dot{Q}}{A_i \Delta T} \quad \text{and} \quad U_o = \frac{\dot{Q}}{A_o \Delta T}$$

We then have

$$\frac{1}{U_i A_i} = \frac{1}{U_o A_o} = R_{total} = \frac{1}{h_i A_i} + \frac{\ln(D_o / D_i)}{2\pi L K} + \frac{1}{h_o A_o}$$

Over time the thermal resistance of the heat exchanger will increase due to surface deposits from the fluids. This effect is accounted for by a fouling factor R_f. The overall heat transfer coefficients become

$$\frac{1}{U_i A_i} = \frac{1}{U_o A_o} = \frac{R_{fi}}{A_i} + \frac{1}{h_i A_i} + \frac{\ln(D_o / D_i)}{2\pi L K} + \frac{1}{h_o A_o} + \frac{R_{fo}}{A_o}$$

Example: The inside tube in a heat exchanger has an inner diameter of 0.1m and an outer diameter of 0.12m. The thermal conductivity of the tube material is k=13W/m°C. The convection heat transfer coefficients are h_i=600W/m²°C and h_o=800W/m²°C. The fouling factors are R_{fi}=0.006m²°C/W and R_{fo}=0.003m²°C/W. Find the inner surface overall heat transfer coefficient.

We can use

$$\frac{1}{U_i (\pi D_i L)} = \frac{R_{fi}}{\pi D_i L} + \frac{1}{h_i (\pi D_i L)} + \frac{\ln(D_o / D_i)}{2\pi L K} + \frac{1}{h_o (\pi D_o L)} + \frac{R_{fo}}{\pi D_o L}$$

Then

$$\frac{1}{U_i} = R_{fi} + \frac{1}{h_i} + \frac{D_i \ln(D_o / D_i)}{2K} + \frac{D_i}{h_o D_o} + \frac{R_{fo} D_i}{D_o}$$

$$\frac{1}{U_i} = 0.006 m^2 °C / W + \frac{1}{600 W / m^2 °C} + \frac{(0.1m) \ln(0.12m / 0.1m)}{2(13 W / m°C)}$$

$$+ \frac{0.1m}{800 W / m^2 °C (0.12m)} + \frac{0.003 m^2 °C / W (0.1m)}{0.12m}$$

$$\frac{1}{U_i} = 0.0119 m^2 °C / W$$

Therefore, $U_i = 84 W / m^2 °C$

To account for the variation in temperature along the length of the pipe, we express the heat transfer rate as

$$\dot{Q} = U A \Delta T_{LM}$$

where U is the overall heat transfer coefficient, and ΔT_{LM} is the log mean temperature difference. For a <u>parallel flow heat exchanger</u> as shown in Figure 13-5,

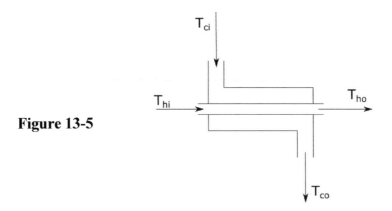

Figure 13-5

the log mean temperature difference is

$$\Delta T_{LM} = \frac{(T_{ho} - T_{co}) - (T_{hi} - T_{ci})}{\ln[(T_{ho} - T_{co})/(T_{hi} - T_{ci})]}$$

For a <u>counter flow heat exchanger</u> as shown in Figure 13-6,

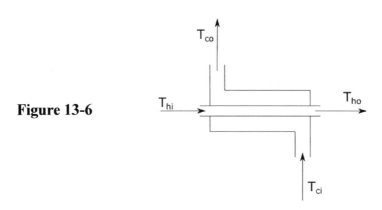

Figure 13-6

the log mean temperature difference is

$$\Delta T_{LM} = \frac{(T_{ho} - T_{ci}) - (T_{hi} - T_{co})}{\ln[(T_{ho} - T_{ci})/(T_{hi} - T_{co})]}$$

The heat transfer rate is related to the mass flow rate as

$$\dot{Q} = \dot{m}_c C_{Pc}(T_{co} - T_{ci}) \quad \text{or} \quad \dot{Q} = \dot{m}_h C_{Ph}(T_{hi} - T_{ho})$$

where C_P is the specific heat. For the special case where the heat exchanger is a condenser or boiler, one of the fluids undergoes a phase change. In this case

$$\dot{Q} = \dot{m}h_{fg} \quad \text{where } h_{fg} \text{ is the specific enthalpy}$$

.

Example: Condenser

Saturated steam at 212°F enters a condenser and exits as saturated liquid at 212°F with a flow rate of 2lbm/s. Cooling water is available at 60°F and must exit at a temperature no higher than 130°F. Find the required mass flow rate of the cooling water if C_p=1BTU/lbm°F.

Using steam table A-2, the heat transfer rate out of the steam is

$$\dot{Q} = \dot{m}_s h_{fg} = 2lbm/s(970.3BTU/lbm) = 1941BTU/s$$

This must equal the heat transfer rate into the cooling water.

$$\dot{Q} = \dot{m}_c C_p (T_{co} - T_{ci})$$

$$\dot{m}_c = \frac{\dot{Q}}{C_P(T_{co} - T_{ci})} = \frac{1941BTU/s}{1BTU/lbm°F(130°F - 60°F)} = 27.7lbm/s$$

Example: Oil cooler

Oil flows into a heat exchanger at 2lbm/s at 200°F. Cooling water is available at 4lbm/s at 80°F. The total heat transfer coefficient for the heat exchanger is 0.3BTU/sft²°F. If the oil outflow temperature needs to be 120°F, find the required heat transfer area if C_p=0.5BTU/lbm°F for oil and C_p=1.0BTU/lbm°F for water.

First we need to find the exit temperature of the water. The heat transferred out of the oil is transferred into the water.

$$\dot{m}_c C_{pc}(T_{co} - T_{ci}) = \dot{m}_h C_{ph}(T_{hi} - T_{ho})$$

Solving for T_{co} gives

$$T_{co} = T_{ci} + \frac{\dot{m}_h C_{ph}(T_{hi} - T_{ho})}{\dot{m}_c C_{pc}}$$

$$T_{co} = 80°F + \frac{2lbm/s(0.5BTU/lbm°F)(200°F - 120°F)}{4lbm/s(1BTU/lbm°F)} = 100°F$$

For a parallel flow heat exchanger,

$$(T_{hi} - T_{ci}) = 200°F - 80°F = 120°F$$
$$(T_{ho} - T_{co}) = 120°F - 100°F = 20°F$$

$$\Delta T_{LM} = \frac{20°F - 120°F}{\ln(20°F/120°F)} = 55.8°F$$

$$A = \frac{\dot{Q}}{U\Delta T_{LM}} = \frac{\dot{m}_c C_{pc}(T_{co} - T_{ci})}{U\Delta T_{LM}} = \frac{4lbm/s(1.0BTU/lbm°F)(100°F - 80°F)}{0.3BTU/sft^2°F(55.8°F)} = 4.78 ft^2$$

For a counter flow heat exchanger,

$$(T_{hi} - T_{co}) = 200°F - 100°F = 100°F$$

$$(T_{ho} - T_{ci}) = 120°F - 80°F = 40°F$$

$$\Delta T_{LM} = \frac{40°F - 100°F}{\ln(40°F/100°F)} = 65.5°F$$

$$A = \frac{\dot{Q}}{U\Delta T_{LM}} = \frac{\dot{m}_c C_{pc}(T_{co} - T_{ci})}{U\Delta T_{LM}} = \frac{4lbm/s(1.0BTU/lbm°F)(100°F - 80°F)}{0.3BTU/sft^2°F(65.5°F)} = 4.07 ft^2$$

Note that the counter flow heat exchanger requires less heat transfer area than the parallel flow heat exchanger. This is usually the case.

13.4 Radiation

We will consider radiation heat transfer between multiple surfaces. The relative orientation of a surface is accounted for with the shape factor F_{ij} where

F_{ij}=fraction of radiation leaving surface i that strikes surface j $(0 \le F_{ij} \le 1)$

F_{ii}=fraction of radiation leaving surface i that strikes itself

A flat or convex surface cannot "see" itself. Therefore, F_{ii}=0 for this case.

The reciprocity rule requires that

$$A_i F_{ij} = A_j F_{ji}$$

where A_i is the area of surface i, and A_j is the area of surface j.

The summation rule requires that

$$\sum_{j=1}^{n} F_{ij} = 1$$

Consider the case of two surfaces forming an enclosure as shown in Figure 13-7.

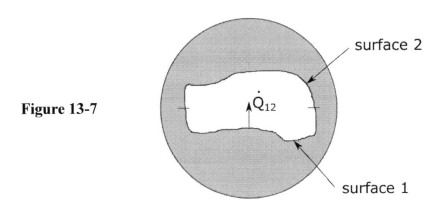

Figure 13-7

For this case the heat transfer rate is

$$\dot{Q}_{12} = \frac{\sigma(T_1^4 - T_2^4)}{\dfrac{1-\varepsilon_1}{A_1\varepsilon_1} + \dfrac{1}{A_1 F_{12}} + \dfrac{1-\varepsilon_2}{A_2\varepsilon_2}}$$

Example: A 3m length of pipe with an outer diameter of 0.1m has a surface temperature of 90°C. It passes through a cylindrical enclosure with an inner diameter of 0.3m that has a surface temperature of 10°C as shown in Figure 13-8. If the emissivity of both surfaces is 0.9, find the radiation heat transfer rate from the pipe to the enclosure.

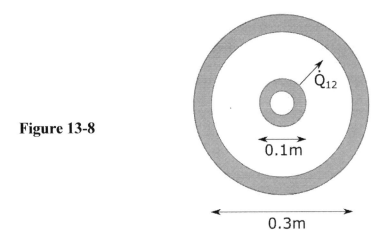

Figure 13-8

0.1m

0.3m

We have

$A_1 = \pi(0.1m)(3m) = 0.942m^2$
$A_2 = \pi(0.3m)(3m) = 2.827m^2$
$F_{12} = 1$
$T_1 = 363°K$
$T_2 = 283°K$

Then

$$\dot{Q}_{12} = \frac{5.67 \times 10^{-8} W / m^2 {}^\circ K[(363^\circ K)^4 - (283^\circ K)^4]}{\dfrac{1-0.9}{(0.942m^2)(0.9)} + \dfrac{1}{(0.942m^2)(1.0)} + \dfrac{1-0.9}{(2.827m^2)(0.9)}} = 509W$$

For radiation heat transfer where A_2 is much larger than A_1 and surface 1 is flat or convex (i.e., $F_{12}=0$), the equation above simplifies to

$$\dot{Q} = \varepsilon \sigma A (T_{s1}^4 - T_{s2}^4)$$

Example: Cooling of a pie
When a pie is removed from an oven, it has a temperature of 200°F. It is placed in a room at a temperature of 72°F. Find the rate of heat transfer from the pie due to radiation alone if it has a surface area of 0.8ft².

If the room is painted white, it has an emissivity of about 0.95. We will assume that the pie has a similar value of emissivity. Then

$$\dot{Q} = 0.95 \left(0.1714x10^{-8} \frac{BTU}{hrft^2 {}^\circ R^4} \right) (0.8 ft^2)[(660^\circ R)^4 - (532^\circ R)^4] = 143 BTU / hr$$

As the pie cools, this rate will decrease because T_{s1} gets smaller.

If heat is also lost by convection during this process, the two heat transfer rates would have to be added together to get the total.

14. ENGINEERING ECONOMICS

14.1 Time-Value of Money Analysis

An economic analysis must account for the effect of time on the value of money. If we need to borrow money to complete a project, the cost of borrowing the money depends on how long we take to repay the loan. The cost of time is determined by the interest rate, which is normally expressed as some percentage of the loan per year. If the loan is repaid in increments over time, determining the cost may require some elaborate calculations.

To start, let's consider a savings account in a bank with an interest rate i=5% per year. If we deposit $100 today (present worth), it will accumulate $5 interest for a total of $105 one year from now (future worth). During the second year, interest will be paid on $105. We can generalize this process using the following notation

     i=interest rate
     n=number of years
     P=present worth
     F=future worth
     A=annual payments

Suppose we put P into a savings account with an annual interest rate of i. The interest earned in the first year is iP. Therefore, the total in the account after the first year is

$$F_1 = P + iP$$

The total after the second year will be

$$F_2 = P + iP + i(P + iP) = P(1 + 2i + i^2) = P(1 + i)^2$$

The total after the third year will be

$$F_3 = P + iP + i(P + iP) + i[P + iP + i(P + iP)] = P(1 + i)^3$$

The total after n years will be

$$F = P(1 + i)^n$$

The term $(1+i)^n$ is called the single payment compound amount factor. Factors of this sort are typically represented by the following notation

$$(Y / X, i\%, n)$$

which is used to find Y given X with an interest rate i for n years. The previous equation for a savings account can be written as

$$F = P(F/P, i\%, n) \quad \text{where} \quad (F/P, i\%, n) = (1+i)^n$$

If a fixed amount is deposited *every* year, then we can calculate the future worth F from an annual payment A over n years as

$$F = P(F/A, i\%, n) \quad \text{where} \quad (F/A, i\%, n) = \frac{[1-(1+i)^{-n}]}{i}$$

The factors for various scenarios are given below.

14.1.1 Compound amount factor (single payment) $(F/P, i\%, n) = (1+i)^n$

This factor gives the future worth F after n years of an account with a present worth P.

Example: Periodically, over a period of ten years, a woman receives dividends on her stocks and deposits them into a savings account earning 5% interest. The deposits are as follows: year one - $1,000; year four - $1,200; year seven - $2,000. How much is in her account after ten years?

Using the formula gives

$$F = \$1,000(F/P, 5\%, 10) + \$1,200(F/P, 5\%, 7) + \$2,000(F/P, 5\%, 4)$$
$$F = \$1,000(1+0.05)^{10} + \$1,200(1+0.05)^7 + \$2,000(1+0.05)^4 = \$5,749$$

Values for this and other factors are available in tables. Such a table for an interest rate of 5% is given ion Table 14-1. From the F/P column (fourth column) in the table, we get

$$F = \$1,000(1.629) + \$1,200(1.407) + \$2,000(1.216) = \$5,749$$

14.1.2 Compound amount factor (uniform series of payments) $(F/A, i\%, n) = \frac{(1+i)^n - 1}{i}$

This factor gives the future worth F that will be accumulated from annual payments A over a period of n years.

Example: A couple is saving for a down payment on a house. They plan to deposit their tax refund each year. If the interest rate is 5%, and they estimate that they will deposit $4000 per year, how much will be accumulated in 5 years?

From the F/A column in Table 14-1, we get

$$F = \$4,000(F/A, 5\%, 5) = \$4,000(5.526) = \$22,104$$

Table 14-1 Interest Rate Table for i=5%

n	P/F	P/A	F/P	F/A	A/P	A/F
1	0.9524	0.952	1.05	1	1.05	1
2	0.907	1.859	1.102	2.05	0.5378	0.4878
3	0.8638	2.723	1.158	3.152	0.3672	0.3172
4	0.8227	3.546	1.216	4.31	0.282	0.232
5	0.7835	4.329	1.276	5.526	0.231	0.181
6	0.7462	5.076	1.34	6.802	0.197	0.147
7	0.7107	5.786	1.407	8.142	0.1728	0.1228
8	0.6768	6.463	1.477	9.549	0.1547	0.1047
9	0.6446	7.108	1.551	11.027	0.1407	0.0907
10	0.6139	7.722	1.629	12.578	0.1295	0.0795
11	0.5847	8.306	1.71	14.207	0.1204	0.0704
12	0.5568	8.863	1.796	15.917	0.1128	0.0628
13	0.5303	9.394	1.886	17.713	0.1065	0.0565
14	0.5051	9.899	1.98	19.599	0.101	0.051
15	0.481	10.38	2.079	21.579	0.0963	0.0463
16	0.4581	10.838	2.183	23.657	0.0923	0.0423
17	0.4363	11.274	2.292	25.84	0.0887	0.0387
18	0.4155	11.69	2.407	28.132	0.0855	0.0355
19	0.3957	12.085	2.527	30.539	0.0827	0.0327
20	0.3769	12.462	2.653	33.066	0.0802	0.0302
21	0.3589	12.821	2.786	35.719	0.078	0.028
22	0.3419	13.163	2.925	38.505	0.076	0.026
23	0.3256	13.489	3.072	41.43	0.0741	0.0241
24	0.3101	13.799	3.225	44.502	0.0725	0.0225
25	0.2953	14.094	3.386	47.727	0.071	0.021
26	0.2812	14.375	3.556	51.113	0.0696	0.0196
27	0.2678	14.643	3.733	54.669	0.0683	0.0183
28	0.2551	14.898	3.92	58.402	0.0671	0.0171
29	0.2429	15.141	4.116	62.323	0.066	0.016
30	0.2314	15.372	4.322	66.439	0.0651	0.0151

14.1.3 Present worth factor (single payment) $(P/F, i\%, n) = \dfrac{1}{(1+i)^n}$

This factor gives the required present worth P to produce a future worth F after n years.

Example: A man wants to buy a bond for his grandchild that can be redeemed for at least $100,000 in 20 years. If the interest rate is 5%, what value of bond must be purchased today?

From the P/F column in Table 14-1, we get

$$P = \$100,000(P/F, 5\%, 20) = \$100,000(0.3769) = \$37,690$$

14.1.4 Present Worth factor (uniform series of payments) $(P/A, i\%, n) = \dfrac{1-(1+i)^{-n}}{i}$

This factor gives the present worth P required to provide annual payment A for n years.

Example: A woman planning for retirement decides that she will need $50,000 per year for 20 years after she retires. How much will she need in her account when she retires if the interest rate is 5%?

From the P/A column in Table 14-1, we get

$$P = \$50,000(P/A, 5\%, 20) = \$50,000(12.462) = \$623,100$$

14.1.5 Sinking fund factor (uniform series of payments) $(A/F, i\%, n) = \dfrac{i}{(1+i)^n - 1}$

This factor gives the required annual payment A needed to accumulate a future worth F after n years.

Example: A couple saving for a down payment on a house would like to have $20,000 accumulated after five years. They plan to deposit their work bonus each year. If the interest rate is 5%, how much would they have to deposit each year?

From the A/F column in Table 14-1, we get

$$A = \$20,000(A/F, 5\%, 5) = \$20,000(0.181) = \$3,620$$

14.1.6 Capital recovery factor (uniform series of payments) $(A/P, i\%, n) = \dfrac{i}{1-(1+i)^{-n}}$

This factor gives the annual payment A needed to deplete an account with present worth P over n years.

Example: A man sets up a trust for his children that contains $200,000. If the interest rate is 5%, how much can the children withdraw each year if they want the money to last 20 years?

From the A/P column in Table 14-1, we get

$$A = \$200,000(A/P, 5\%, 20) = \$200,000(0.0802) = \$16,040$$

14.2 Economic Alternative Comparison

A common method for comparing economic alternatives is to calculate the *equivalent uniform annual cost* (EUAC). This is typically done for a piece of equipment with an initial cost P, salvage value of SV, and an annual operating cost AOC. Then,

$$EUAC = P(A/P, i\%, n) - SV(A/F, i\%, n) + AOC$$

Example: A company must decide between buying machine A that costs \$100,000, has an annual operating cost of \$5,000, and has a salvage value of \$15,000 after 10 years versus buying machine B that costs \$80,000, has an annual operating cost of \$6,200, and has a salvage value of \$5,000 after 10 years. Compare these alternatives using EAUC if the interest rate is 5%.

First, consider machine A.

$$EUAC = \$100,000(A/P, 5\%, 10) - \$15,000(A/F, 5\%, 10) + \$5,000$$
$$EUAC = \$100,000(0.1295) - \$15,000(0.0795) + \$5,000 = \$16,758$$

Next, consider machine B.

$$EUAC = \$80,000(A/P, 5\%, 10) - \$5,000(A/F, 5\%, 10) + \$6,500$$
$$EUAC = \$80,000(0.1295) - \$5,000(0.0795) + \$6,200 = \$16,163$$

Machine B has a lower equivalent uniform annual cost than machine A, which makes machine B the better economic value.

14.3 Depreciation

Depreciation is a means to account for the loss in values of assets, such as machines, buildings, etc, over time. Depreciation of assets allows the owner to take a deduction in income taxes. Two methods for accounting for depreciation are described below.

14.3.1 Straight-line method

In the straight line method, an asset's book value BV (the purchase value minus the accumulated depreciation) depreciates linearly over time; i.e.,

$$D = \frac{PV - SV}{n}$$

where D is the annual depreciation charge, PV is the purchase value, SV is the salvage value, and n is the number of years of total life of the asset. The book value of the asset after j years is

$$BV = PV - jD = PV - j\left(\frac{PV - SV}{n}\right)$$

14.3.2 Modified accelerated cost recovery system method

The 1986 Tax Reform Act instituted the modified accelerated cost recovery system method which dictates the depreciation charge in a given year j as

$$D_j = d(PV)$$

where the value d is given in Table 14-2. The book value is given as

$$BV = PV - \sum_{j=1}^{n} D_j$$

Example: Depreciation of a machine

A piece of machinery has a purchase value of $50,000 and a salvage value of $10,000 after 10 years. Calculate the book value after 5 years using the above methods.

Straight-Line Method

$$BV_{SL} = PV - j\left(\frac{PV - SV}{n}\right) = \$50,000 - 5\left(\frac{\$50,000 - \$10,000}{10}\right) = \$30,000$$

MACRS Method

$$BV_{MACRS} = \$50,000 - \$50,000(0.10 + 0.18 + 0.144 + 0.1152 + 0.0922) = \$18,430$$

Table 14-2 MARCS Recovery Rates

Year	3-year	5-year	7-year	10-year
1	33.33%	20.00%	14.29%	10.00%
2	44.45	32	24.49	18
3	14.81	19.2	17.49	14.4
4	7.41	11.52	12.49	11.52
5		11.52	8.93	9.22
6		5.76	8.92	7.37
7			8.93	6.55
8			4.46	6.55
9				6.56
10				6.55
11				3.28

APPENDIX A THERMODYNAMIC PROPERTIES TABLES

Table A-1 Properties of Saturated Steam for Increments in Temperature

T(°F)	p_{sat}(psi)	v_f(ft³/lbm)	v_g(ft³/lbm)	h_f(BTU/lbm)	h_g(BTU/lbm)	s_f(BTU/lbm°R)	s_g(BTU/lbm°R)
32	0.08866	0.01600	3302	0.001	1075.4	0	2.1869
40	0.12173	0.01602	2443.3	8.0377	1079.4	0.016215	2.1604
50	0.17814	0.016024	1702.8	18.078	1083.8	0.03611	2.1271
60	0.2564	0.016035	1206.0	28.096	1088.2	0.055576	2.0954
70	0.36336	0.016052	867.08	38.101	1092.5	0.074644	2.0653
80	0.50747	0.016074	632.39	48.097	1096.8	0.09334	2.0366
90	0.69904	0.0161	467.40	58.088	1101.1	0.11168	2.0093
100	0.95051	0.016131	349.83	68.078	1105.4	0.12969	1.9832
110	1.2767	0.016166	264.96	78.069	1109.7	0.14738	1.9583
120	1.695	0.016205	202.95	88.062	1114	0.16477	1.9346
130	2.2259	0.016247	157.09	98.059	1118.2	0.18187	1.9118
140	2.893	0.016293	122.82	108.06	1122.3	0.19868	1.8901
150	3.7232	0.016342	96.927	118.07	1126.5	0.21523	1.8693
160	4.7472	0.016394	77.184	128.08	1130.6	0.23152	1.8493
170	5.9998	0.016449	61.981	138.11	1134.6	0.24756	1.8302
180	7.5195	0.016508	50.168	148.14	1138.6	0.26336	1.8118
190	9.3496	0.016569	40.917	158.18	1142.6	0.27893	1.7942
200	11.538	0.016633	33.609	168.24	1146.5	0.29429	1.7772
210	14.136	0.016701	27.795	178.31	1150.3	0.30943	1.7609
220	17.201	0.016771	23.133	188.4	1154.1	0.32436	1.7451
230	20.795	0.016845	19.371	198.51	1157.7	0.33911	1.7299
240	24.986	0.016921	16.314	208.63	1161.3	0.35366	1.7153
250	29.844	0.017001	13.815	218.78	1164.8	0.36804	1.7011
260	35.447	0.017084	11.759	228.95	1168.2	0.38225	1.6874
270	41.878	0.01717	10.058	239.14	1171.5	0.39629	1.6741
280	49.222	0.017259	8.6430	249.37	1174.7	0.41017	1.6612
290	57.574	0.017352	7.4599	259.62	1177.8	0.42391	1.6487
300	67.029	0.017448	6.4658	269.91	1180.7	0.4375	1.6365
310	77.691	0.017548	5.6262	280.23	1183.6	0.45096	1.6246
320	89.667	0.017652	4.9142	290.6	1186.3	0.46428	1.6131
330	103.07	0.01776	4.3074	301	1188.8	0.47749	1.6018
340	118.02	0.017872	3.7883	311.45	1191.3	0.49058	1.5908
350	134.63	0.017988	3.3425	321.95	1193.5	0.50355	1.58
		$v_{fg} = v_g - v_f$		$h_{fg} = h_g - h_f$		$s_{fg} = s_g - s_f$	

Table A-1 Properties of Saturated Steam for Increments in Temperature

T(°F)	p_{sat}(psi)	v_f(ft³/lbm)	v_g(ft³/lbm)	h_f(BTU/lbm)	h_g(BTU/lbm)	s_f(BTU/lbm°R)	s_g(BTU/lbm°R)
360	153.03	0.018108	2.9580	332.5	1195.6	0.51643	1.5694
370	173.36	0.018233	2.6252	343.11	1197.5	0.52921	1.5591
380	195.74	0.018363	2.3361	353.77	1199.3	0.5419	1.5489
390	220.33	0.018498	2.0842	364.5	1200.9	0.5545	1.5388
400	247.26	0.018639	1.8638	375.3	1202.2	0.56703	1.5289
410	276.68	0.018785	1.6706	386.17	1203.4	0.57948	1.5192
420	308.76	0.018938	1.5006	397.12	1204.4	0.59187	1.5096
430	343.64	0.019097	1.3505	408.15	1205.1	0.6042	1.5
440	381.48	0.019263	1.2177	419.27	1205.6	0.61648	1.4905
450	422.46	0.019437	1.0999	430.49	1205.9	0.62872	1.4811
460	466.75	0.019619	0.99502	441.81	1205.9	0.64092	1.4718
470	514.52	0.019809	0.90155	453.24	1205.7	0.65309	1.4624
480	565.95	0.02001	0.81793	464.78	1205.1	0.66524	1.4531
490	621.23	0.02022	0.74294	476.46	1204.3	0.67738	1.4438
500	680.55	0.020442	0.67554	488.27	1203.1	0.68952	1.4344
510	744.11	0.020675	0.61485	500.23	1201.6	0.70167	1.425
520	812.1	0.020923	0.56007	512.35	1199.8	0.71384	1.4155
530	884.74	0.021185	0.51049	524.65	1197.5	0.72604	1.4059
540	962.24	0.021464	0.46553	537.14	1194.8	0.73829	1.3962
550	1044.8	0.021761	0.42466	549.84	1191.7	0.7506	1.3863
560	1132.7	0.02208	0.38742	562.77	1188	0.76299	1.3762
570	1226.2	0.022422	0.35341	575.96	1183.8	0.77549	1.3658
580	1325.5	0.022792	0.32227	589.44	1179	0.78812	1.3552
590	1430.8	0.023193	0.29369	603.25	1173.6	0.8009	1.3443
600	1542.5	0.023631	0.26739	617.42	1167.4	0.81388	1.3329
		$v_{fg} = v_g - v_f$		$h_{fg} = h_g - h_f$		$s_{fg} = s_g - s_f$	

†The properties in this table were calculated using the NIST program REFPROPMINI.

Table A-2 Properties of Saturated Steam for Increments in Pressure

p_{sat}(psi)	T(°F)	v_f(ft³/lbm)	v_g(ft³/lbm)	h_f(BTU/lbm)	h_g(BTU/lbm)	s_f(BTU/lbm°R)	s_g(BTU/lbm°R)
14.7	212	0.016715	26.800	180.15	1150.5	0.31212	1.7557
15	212.99	0.016722	26.294	181.33	1151.4	0.31391	1.7561
20	227.92	0.016829	20.091	196.4	1157	0.33605	1.7331
25	240.03	0.016922	16.305	208.66	1161.3	0.35371	1.7152
30	250.3	0.017003	13.747	219.08	1164.9	0.36847	1.7007
35	259.25	0.017077	11.900	228.18	1168	0.38119	1.6884
40	267.22	0.017146	10.500	236.3	1170.6	0.3924	1.6777
45	274.41	0.017209	9.4020	243.65	1172.9	0.40243	1.6684
50	280.99	0.017268	8.5172	250.38	1175	0.41153	1.6599
55	287.05	0.017324	7.7876	256.59	1176.9	0.41987	1.6523
60	292.68	0.017378	7.1762	262.38	1178.6	0.42757	1.6454
65	297.95	0.017429	6.6556	267.8	1180.2	0.43473	1.639
70	302.91	0.017477	6.2069	272.91	1181.6	0.44143	1.633
75	307.58	0.017524	5.8163	277.74	1182.9	0.44772	1.6275
80	312.02	0.017569	5.4732	282.32	1184.1	0.45366	1.6223
85	316.24	0.017613	5.1688	286.69	1185.3	0.45928	1.6174
90	320.26	0.017655	4.8969	290.87	1186.3	0.46463	1.6128
95	324.11	0.017696	4.6529	294.87	1187.3	0.46973	1.6084
100	327.81	0.017736	4.4326	298.71	1188.3	0.4746	1.6043
110	334.77	0.017813	4.0499	305.98	1190	0.48375	1.5965
120	341.25	0.017886	3.7288	312.76	1191.5	0.49221	1.5894
130	347.32	0.017956	3.4557	319.13	1192.9	0.50008	1.5829
140	353.03	0.018024	3.2201	325.14	1194.2	0.50746	1.5768
150	358.42	0.018088	3.0150	330.83	1195.3	0.5144	1.5711
160	363.54	0.018151	2.8346	336.25	1196.3	0.52097	1.5657
170	368.41	0.018213	2.6748	341.42	1197.2	0.52719	1.5607
180	373.07	0.018272	2.5322	346.37	1198.1	0.53311	1.5559
190	377.52	0.01833	2.4040	351.13	1198.9	0.53876	1.5514
200	381.8	0.018387	2.2882	355.7	1199.6	0.54417	1.547
		$v_{fg} = v_g - v_f$		$h_{fg} = h_g - h_f$		$s_{fg} = s_g - s_f$	

Table A-2 Properties of Saturated Steam for Increments in Pressure

p$_{sat}$(psi)	T(°F)	v$_f$(ft³/lbm)	v$_g$(ft³/lbm)	h$_f$(BTU/lbm)	h$_g$(BTU/lbm)	s$_f$(BTU/lbm°R)	s$_g$(BTU/lbm°R)
250	400.97	0.018653	1.8440	376.35	1202.4	0.56824	1.528
300	417.35	0.018897	1.5435	394.21	1204.1	0.58859	1.5121
350	431.74	0.019126	1.3263	410.08	1205.2	0.60634	1.4984
400	444.62	0.019342	1.1616	424.44	1205.8	0.62213	1.4862
450	456.31	0.019551	1.0324	437.62	1206	0.63642	1.4752
500	467.04	0.019752	0.92816	449.84	1205.8	0.64949	1.4652
600	486.24	0.02014	0.77018	472.06	1204.6	0.67282	1.4473
700	503.13	0.020513	0.65587	492	1202.7	0.69332	1.4315
800	518.27	0.020879	0.56918	510.24	1200.1	0.71173	1.4171
900	532.02	0.02124	0.50105	527.16	1197	0.72851	1.4039
1000	544.65	0.0216	0.44605	543.02	1193.4	0.744	1.3916
1100	556.35	0.021961	0.40062	558.02	1189.4	0.75846	1.3799
1200	567.26	0.022325	0.36244	572.32	1185	0.77205	1.3687
1300	577.49	0.022696	0.32983	586.03	1180.3	0.78493	1.3579
1400	587.14	0.023074	0.30163	599.26	1175.2	0.79722	1.3474
1500	596.26	0.023462	0.27697	612.08	1169.8	0.809	1.3372
1600	604.93	0.023862	0.25519	624.56	1164	0.82036	1.3271
1700	613.18	0.024277	0.23577	636.77	1157.9	0.83137	1.3171
1800	621.07	0.024708	0.21832	648.76	1151.4	0.84208	1.3072
1900	628.61	0.02516	0.20252	660.58	1144.5	0.85255	1.2972
2000	635.85	0.025635	0.18813	672.28	1137.2	0.86284	1.2872
		v$_{fg}$ = v$_g$-v$_f$		h$_{fg}$ = h$_g$-h$_f$		s$_{fg}$ = s$_g$-s$_f$	

†The properties in this table were calculated using the NIST program REFPROPMINI.

Table A-3 Properties of Superheated Steam

p(psi)	T(°F)	v(ft^3/lbm)	h(BTU/lbm)	s(BTU/lbm°R)
100	400	4.9358	1228.6	1.6532
100	500	5.5875	1280.1	1.71
100	600	6.2166	1330.3	1.7598
100	700	6.8343	1380.4	1.8049
100	800	7.4455	1430.8	1.8466
100	900	8.0528	1481.7	1.8855
100	1000	8.6573	1533.4	1.9222
100	1100	9.2601	1585.9	1.9569
100	1200	9.8610	1639.2	1.99
100	1300	10.462	1693.3	2.0217
100	1400	11.061	1748.4	2.0522
100	1500	11.660	1804.4	2.0815
100	1600	12.258	1861.2	2.1098
100	1700	12.856	1919	2.1372
100	1800	13.454	1977.6	2.1637
100	1900	14.051	2037.1	2.1894
100	2000	14.649	2097.4	2.2145
200	400	2.3614	1211.7	1.5613
200	500	2.7247	1269.9	1.6254
200	600	3.0586	1323.2	1.6782
200	700	3.3795	1375	1.725
200	800	3.6934	1426.6	1.7676
200	900	4.0030	1478.3	1.8072
200	1000	4.3098	1530.6	1.8443
200	1100	4.6147	1583.5	1.8793
200	1200	4.9181	1637.1	1.9127
200	1300	5.2206	1691.6	1.9445
200	1400	5.5221	1746.9	1.9751
200	1500	5.8231	1803.1	2.0045
200	1600	6.1237	1860.1	2.0329
200	1700	6.4238	1918	2.0603
200	1800	6.7236	1976.7	2.0869
200	1900	7.0235	2036.3	2.1127
200	2000	7.3228	2096.7	2.1378

Table A-3 Properties of Superheated Steam

p(psi)	T(°F)	v(ft³/lbm)	h(BTU/lbm)	s(BTU/lbm°R)
300	500	1.7670	1258.7	1.5716
300	600	2.0046	1315.7	1.6282
300	700	2.2273	1369.5	1.6767
300	800	2.4424	1422.3	1.7204
300	900	2.6529	1474.9	1.7605
300	1000	2.8606	1527.7	1.7981
300	1100	3.0662	1581.1	1.8334
300	1200	3.2704	1635.1	1.867
300	1300	3.4735	1689.9	1.899
300	1400	3.6759	1745.4	1.9297
300	1500	3.8776	1801.8	1.9592
300	1600	4.0790	1858.9	1.9877
300	1700	4.2797	1917	2.0152
300	1800	4.4803	1975.8	2.0418
300	1900	4.6806	2035.5	2.0677
300	2000	4.8807	2096	2.0928
400	500	1.2851	1246.4	1.5298
400	600	1.4765	1307.8	1.5907
400	700	1.6507	1363.8	1.6413
400	800	1.8166	1417.9	1.686
400	900	1.9777	1471.4	1.7269
400	1000	2.1358	1524.9	1.7649
400	1100	2.2918	1578.7	1.8005
400	1200	2.4465	1633.1	1.8343
400	1300	2.6000	1688.1	1.8665
400	1400	2.7528	1743.9	1.8973
400	1500	2.9049	1800.4	1.927
400	1600	3.0565	1857.8	1.9555
400	1700	3.2077	1915.9	1.9831
400	1800	3.3586	1974.9	2.0098
400	1900	3.5093	2034.7	2.0356
400	2000	3.6597	2095.2	2.0608

Table A-3 Properties of Superheated Steam

p(psi)	T(°F)	v(ft³/lbm)	h(BTU/lbm)	s(BTU/lbm°R)
500	500	0.99305	1232.8	1.4938
500	600	1.1587	1299.5	1.5601
500	700	1.3044	1357.9	1.6128
500	800	1.4410	1413.4	1.6588
500	900	1.5725	1467.9	1.7003
500	1000	1.7009	1522	1.7387
500	1100	1.8273	1576.3	1.7747
500	1200	1.9521	1631.1	1.8088
500	1300	2.0759	1686.4	1.8411
500	1400	2.1988	1742.4	1.8721
500	1500	2.3212	1799.1	1.9018
500	1600	2.4430	1856.6	1.9304
500	1700	2.5644	1914.9	1.9581
500	1800	2.6856	1974	1.9848
500	1900	2.8065	2033.9	2.0107
500	2000	2.9271	2094.5	2.0359
600	500	0.79523	1217.3	1.4606
600	600	0.94607	1290.8	1.5336
600	700	1.0732	1351.9	1.5888
600	800	1.1904	1408.9	1.6359
600	900	1.3023	1464.3	1.6782
600	1000	1.4110	1519.1	1.7171
600	1100	1.5175	1573.9	1.7534
600	1200	1.6225	1629	1.7877
600	1300	1.7265	1684.6	1.8202
600	1400	1.8296	1740.9	1.8513
600	1500	1.9320	1797.8	1.8811
600	1600	2.0340	1855.5	1.9098
600	1700	2.1356	1913.9	1.9375
600	1800	2.2369	1973.1	1.9643
600	1900	2.3379	2033.1	1.9903
600	2000	2.4387	2093.8	2.0155

Table A-3 Properties of Superheated Steam

p(psi)	T(°F)	v(ft³/lbm)	h(BTU/lbm)	s(BTU/lbm°R)
800	600	0.67797	1271.7	1.4877
800	700	0.78333	1339.3	1.5487
800	800	0.87681	1399.6	1.5986
800	900	0.96432	1457	1.6425
800	1000	1.0484	1513.2	1.6823
800	1100	1.1302	1569	1.7193
800	1200	1.2105	1624.9	1.7541
800	1300	1.2897	1681.1	1.7869
800	1400	1.3680	1737.8	1.8183
800	1500	1.4457	1795.2	1.8483
800	1600	1.5228	1853.2	1.8772
800	1700	1.5996	1911.9	1.905
800	1800	1.6761	1971.3	1.9319
800	1900	1.7523	2031.5	1.958
800	2000	1.8282	2092.4	1.9833
1000	600	0.51430	1250.1	1.4466
1000	700	0.60846	1325.8	1.5151
1000	800	0.68818	1389.9	1.5681
1000	900	0.76138	1449.6	1.6137
1000	1000	0.83077	1507.2	1.6546
1000	1100	0.89783	1564.1	1.6923
1000	1200	0.96330	1620.8	1.7275
1000	1300	1.0276	1677.6	1.7608
1000	1400	1.0910	1734.8	1.7924
1000	1500	1.1538	1792.5	1.8226
1000	1600	1.2161	1850.9	1.8516
1000	1700	1.2780	1909.9	1.8796
1000	1800	1.3395	1969.5	1.9066
1000	1900	1.4009	2029.9	1.9328
1000	2000	1.4619	2091	1.9581

†The properties in this table were calculated using the NIST program EFPROPMINI.

Table A-4 Properties of Air for Increments in Temperature

T(°R)	h(BTU/lbm)	u(BTU/lbm)	s(BTU/lbm°R)
400	96.0	68.5	0.5289
420	100.6	71.8	0.5405
440	105.3	75.1	0.5515
460	110.0	78.5	0.5621
480	114.7	81.8	0.5722
500	119.4	85.1	0.5820
520	124.1	88.5	0.5914
540	128.9	91.8	0.6004
560	133.6	95.2	0.6092
580	138.4	98.6	0.6176
600	143.2	102.0	0.6258
620	147.9	105.4	0.6338
640	152.7	108.9	0.6415
660	157.5	112.3	0.6489
680	162.4	115.7	0.6562
700	167.2	119.2	0.6633
720	172.0	122.7	0.6702
740	176.9	126.1	0.6769
760	181.8	129.6	0.6834
780	186.6	133.1	0.6898
800	191.5	136.7	0.6960
820	196.4	140.2	0.7021
840	201.3	143.7	0.7081
860	206.3	147.3	0.7139
880	211.2	150.8	0.7196
900	216.1	154.4	0.7252
920	221.1	158.0	0.7307
940	226.1	161.6	0.7361
960	231.0	165.2	0.7413
980	236.0	168.8	0.7465
1000	241.0	172.5	0.7516

Table A-4 Properties of Air for Increments in Temperature

T(°R)	h(BTU/lbm)	u(BTU/lbm)	s(BTU/lbm°R)
1100	266.2	190.8	0.776
1200	291.7	209.4	0.798
1300	317.4	228.3	0.818
1400	343.4	247.4	0.838
1500	369.7	266.9	0.856
1600	396.3	286.6	0.873
1700	423.1	306.5	0.889
1800	450.2	326.8	0.905
1900	477.5	347.3	0.919
2000	505.2	368.0	0.933
2100	533.0	389.0	0.947
2200	561.2	410.3	0.960
2300	589.5	431.8	0.973
2400	618.2	453.6	0.985
2500	647.0	475.6	0.996
2600	676.2	497.9	1.008
2700	705.5	520.4	1.019
2800	735.1	543.1	1.029
2900	765.0	566.1	1.040
3000	795.1	589.4	1.050
3100	825.4	612.8	1.060
3200	855.9	636.5	1.070
3300	886.7	660.4	1.079
3400	917.7	684.6	1.088
3500	948.9	708.9	1.097
3600	980.4	733.5	1.106
3700	1012	758.3	1.115
3800	1044	783.3	1.123
3900	1076	808.6	1.132
4000	1108	834.0	1.140

†The properties in this table were calculated using the formulas in Irvine, T.F. and Liley, P.E., *Steam and Gas Tables with Computer Equations*, Academic Press, Orlando, 1984.

Table A-5 Properties of Saturated Refrigerant 134a for Increments in Temperature

T(°F)	p_{sat}(psi)	v_f(ft³/lbm)	v_g(ft³/lbm)	h_f(BTU/lbm)	h_g(BTU/lbm)	s_f(BTU/lbm°R)	s_g(BTU/lbm°R)
-20	12.898	0.011565	3.444831	5.711	98.81	0.01332	0.2251
-10	16.632	0.011706	2.710909	8.775	100.3	0.0202	0.22372
0	21.171	0.011853	2.157916	11.867	101.78	0.02697	0.22251
10	26.628	0.012007	1.735749	14.989	103.24	0.03367	0.22146
20	33.124	0.012168	1.409423	18.141	104.68	0.04028	0.22053
30	40.784	0.012337	1.154348	21.328	106.09	0.04682	0.21972
40	49.741	0.012515	0.952835	24.55	107.48	0.05329	0.21901
50	60.134	0.012703	0.791954	27.81	108.83	0.0597	0.21838
60	72.105	0.012903	0.662471	31.112	110.15	0.06606	0.21782
70	85.805	0.013116	0.557227	34.459	111.42	0.07238	0.21731
80	101.39	0.013345	0.471054	37.857	112.64	0.07866	0.21684
90	119.01	0.01359	0.399888	41.307	113.81	0.08491	0.21638
100	138.85	0.013856	0.340692	44.817	114.91	0.09115	0.21593
110	161.07	0.014146	0.291112	48.387	115.94	0.09738	0.21546
120	185.86	0.014464	0.249277	52.047	116.88	0.10362	0.21494
130	213.41	0.014817	0.213739	55.777	117.72	0.10988	0.21436
140	243.92	0.015214	0.183321	59.617	118.43	0.11619	0.21368
150	277.61	0.015666	0.157092	63.567	118.99	0.12257	0.21285
160	314.73	0.016191	0.13428	67.667	119.35	0.12906	0.21182
170	355.53	0.016815	0.114235	71.947	119.46	0.1357	0.2105
180	400.34	0.017588	0.096376	76.467	119.22	0.14261	0.20875
190	449.52	0.018603	0.080115	81.357	118.44	0.14993	0.2063
200	503.59	0.020096	0.064662	86.877	116.72	0.15807	0.20256

†The properties in this table were calculated using the NIST program REFPROPMINI.

Table A-6 Properties of Saturated Refrigerant 134a for Increments in Pressure

p$_{sat}$(psi)	T(°F)	v$_f$(ft^3/lbm)	v$_g$(ft^3/lbm)	h$_f$(BTU/lbm)	h$_g$(BTU/lbm)	s$_f$(BTU/lbm°R)	s$_g$(BTU/lbm°R)
20	-2.4046	0.011817	2.277386	10.887	101.39	0.0248	0.2228
40	29.047	0.01232	1.176111	20.788	105.92	0.04565	0.21981
60	49.88	0.012701	0.793714	27.537	108.78	0.05907	0.2184
80	65.922	0.013028	0.597586	32.855	110.87	0.06926	0.21753
100	79.159	0.013325	0.477669	37.333	112.5	0.07758	0.21689
120	90.526	0.013604	0.396495	41.253	113.83	0.08469	0.21637
140	100.55	0.013871	0.337769	44.773	114.93	0.09094	0.21591
160	109.54	0.014132	0.293204	47.993	115.85	0.09654	0.21549
180	117.73	0.014389	0.258178	50.973	116.64	0.10165	0.21507
200	125.26	0.014645	0.229874	53.763	117.3	0.10636	0.21466
220	132.25	0.014902	0.206492	56.393	117.85	0.11074	0.21423
240	138.77	0.015163	0.186822	58.903	118.31	0.11486	0.21378
260	144.89	0.015428	0.170013	61.303	118.69	0.11875	0.21331
280	150.67	0.015699	0.155458	63.603	118.98	0.12245	0.2128
300	156.15	0.015978	0.142712	65.833	119.2	0.12599	0.21226
320	161.35	0.016268	0.131434	67.993	119.35	0.12939	0.21167
340	166.3	0.01657	0.121361	70.103	119.42	0.13267	0.21104
360	171.04	0.016888	0.112285	72.163	119.42	0.13586	0.21035
380	175.58	0.017224	0.104043	74.193	119.34	0.13896	0.2096
400	179.93	0.017582	0.096497	76.203	119.18	0.14201	0.20877
420	184.11	0.017968	0.089534	78.183	118.94	0.145	0.20786
440	188.14	0.018388	0.083056	80.173	118.6	0.14797	0.20684
460	192.01	0.018852	0.07697	82.163	118.15	0.15093	0.20569
480	195.75	0.019375	0.0712	84.183	117.58	0.15392	0.20439
500	199.37	0.019977	0.065643	86.263	116.83	0.15696	0.20287

†The properties in this table were calculated using the NIST program REFPROPMINI.

Table A-7 Properties of Superheated Refrigerant 134a

p(psi)	T(°F)	v(ft³/lbm)	h(BTU/lbm)	s(BTU/lbm°R)
20	0	2.292211	101.91	0.224
20	20	2.413011	105.82	0.23234
20	40	2.530621	109.77	0.24041
20	60	2.646133	113.78	0.24828
20	80	2.759991	117.86	0.25598
20	100	2.872573	122.01	0.26353
20	120	2.984184	126.24	0.27095
20	140	3.095017	130.55	0.27826
20	160	3.205128	134.93	0.28546
20	180	3.314661	139.4	0.29255
20	200	3.42372	143.95	0.29956
40	40	1.212606	108.29	0.22468
40	60	1.276829	112.5	0.23295
40	80	1.338921	116.73	0.24094
40	100	1.399463	121.01	0.24871
40	120	1.458853	125.34	0.25631
40	140	1.517336	129.73	0.26376
40	160	1.575051	134.19	0.27108
40	180	1.632173	138.72	0.27828
40	200	1.688761	143.33	0.28537
40	220	1.744927	148.01	0.29235
40	240	1.800731	152.76	0.29925
40	260	1.856183	157.6	0.30606
40	280	1.911425	162.51	0.31279

Table A-7 Properties of Superheated Refrigerant 134a

p(psi)	T(°F)	v(ft³/lbm)	h(BTU/lbm)	s(BTU/lbm°R)
60	60	0.81793	111.1	0.22299
60	80	0.86356	115.53	0.23136
60	100	0.90728	119.95	0.23941
60	120	0.94958	124.4	0.24722
60	140	0.99079	128.89	0.25483
60	160	1.03120	133.43	0.26228
60	180	1.07095	138.03	0.26958
60	200	1.11014	142.69	0.27676
60	220	1.14887	147.42	0.28382
60	240	1.18721	152.22	0.29078
60	260	1.22522	157.09	0.29765
60	280	1.26295	162.04	0.30443
60	300	1.30042	167.05	0.31112
80	80	0.62430	114.22	0.22391
80	100	0.66011	118.83	0.23229
80	120	0.69416	123.41	0.24034
80	140	0.72701	128.01	0.24814
80	160	0.75890	132.64	0.25573
80	180	0.79001	137.31	0.26315
80	200	0.82061	142.04	0.27043
80	220	0.85063	146.82	0.27757
80	240	0.88028	151.67	0.2846
80	260	0.90959	156.58	0.29152
80	280	0.93861	161.56	0.29835
80	300	0.96740	166.61	0.30509
80	320	0.99592	171.73	0.31174

Table A-7 Properties of Superheated Refrigerant 134a

p(psi)	T(°F)	v(ft³/lbm)	h(BTU/lbm)	s(BTU/lbm°R)
100	80	0.47906	112.76	0.21745
100	100	0.51078	117.62	0.22629
100	120	0.54022	122.37	0.23464
100	140	0.56821	127.09	0.24264
100	160	0.59513	131.82	0.2504
100	180	0.62123	136.57	0.25795
100	200	0.64666	141.37	0.26533
100	220	0.67159	146.21	0.27256
100	240	0.69604	151.1	0.27966
100	260	0.72015	156.06	0.28664
100	280	0.74399	161.08	0.29352
100	300	0.76752	166.16	0.3003
100	320	0.79083	171.31	0.30699
120	100	0.41014	116.3	0.22091
120	120	0.43693	121.26	0.22962
120	140	0.46192	126.13	0.23788
120	160	0.48565	130.97	0.24582
120	180	0.50844	135.81	0.25351
120	200	0.53053	140.68	0.261
120	220	0.55206	145.58	0.26832
120	240	0.57313	150.53	0.2755
120	260	0.59379	155.53	0.28255
120	280	0.61414	160.59	0.28948
120	300	0.63420	165.71	0.29631
120	320	0.65402	170.89	0.30304

Table A-7 Properties of Superheated Refrigerant 134a

p(psi)	T(°F)	v(ft³/lbm)	h(BTU/lbm)	s(BTU/lbm°R)
160	120	0.30578	118.77	0.22066
160	140	0.32774	124.04	0.2296
160	160	0.34791	129.16	0.23799
160	180	0.36686	134.21	0.24602
160	200	0.38494	139.24	0.25377
160	220	0.40235	144.29	0.26129
160	240	0.41922	149.35	0.26864
160	260	0.43563	154.45	0.27582
160	280	0.45171	159.59	0.28287
160	300	0.46749	164.78	0.2898
160	320	0.48300	170.03	0.29662
160	340	0.49828	175.34	0.30333
160	360	0.51340	180.7	0.30996
200	140	0.24541	121.65	0.2221
200	160	0.26412	127.16	0.23114
200	180	0.28115	132.48	0.2396
200	200	0.29704	137.72	0.24766
200	220	0.31213	142.92	0.25543
200	240	0.32658	148.12	0.26297
200	260	0.34054	153.33	0.27031
200	280	0.35411	158.56	0.27748
200	300	0.36734	163.84	0.28452
200	320	0.38029	169.15	0.29142
200	340	0.39300	174.52	0.29822
200	360	0.40553	179.94	0.30491

Table A-7 Properties of Superheated Refrigerant 134a

p(psi)	T(°F)	v(ft³/lbm)	h(BTU/lbm)	s(BTU/lbm°R)
300	160	0.14656	120.67	0.21474
300	180	0.16356	127.32	0.22531
300	200	0.17777	133.38	0.23463
300	220	0.19044	139.15	0.24325
300	240	0.20211	144.78	0.25141
300	260	0.21307	150.33	0.25924
300	280	0.22347	155.85	0.2668
300	300	0.23346	161.35	0.27414
300	320	0.24310	166.87	0.28131
300	340	0.25247	172.41	0.28832
300	360	0.26160	177.98	0.2952
400	180	0.09658	119.27	0.20901
400	200	0.11440	127.71	0.222
400	220	0.12746	134.6	0.2323
400	240	0.13853	140.93	0.24148
400	260	0.14844	146.98	0.25001
400	280	0.15756	152.88	0.25809
400	300	0.16611	158.68	0.26584
400	320	0.17423	164.44	0.27332
400	340	0.18201	170.19	0.28059
400	360	0.18951	175.93	0.28769

†The properties in this table were calculated using the NIST program REFPROPMINI.

INDEX